THE ULTIMATE GUIDE TO CLINICAL RESEARCH

A COMPLETE HANDBOOK FOR RESEARCHERS AND PRACTITIONERS

SMITH, TYLER

THE ULTIMATE GUIDE TO CLINICAL RESEARCH

The Ultimate Guide to Clinical Research:
A Complete Handbook for Researchers and Practitioners

Tyler Smith

ISBN: 978-965125-00-7

CONTENTS

1

INTRODUCTION TO CLINICAL RESEARCH

CHAPTER I

INTRODUCTION TO CLINICAL RESEARCH

In this chapter, we will provide an overview of clinical research and its significance in advancing medical knowledge and improving patient care. We will explore the basic principles and goals of clinical research, highlighting the key differences between clinical studies and other types of scientific investigations. By the end of this chapter, you will have a clear understanding of the fundamental concepts that underpin clinical research.

SECTION 1: WHAT IS CLINICAL RESEARCH?

What is Clinical Research?

In this section, we will provide a comprehensive introduction to clinical research. We will explain that clinical research is a branch of scientific inquiry that aims to investigate the safety, efficacy, and effectiveness of medical interventions, such as drugs, treatments, medical devices, and behavioral interventions. Clinical research involves systematically gathering data from human participants to generate evidence that guides medical decision-making and improves patient outcomes.

We will discuss the objectives of clinical research, which include:

- Assessing the safety and tolerability of new treatments or interventions.
- Determining the effectiveness of interventions in treating specific medical conditions.
- Comparing the efficacy of different interventions to identify the most suitable treatment options.
- Understanding the natural history and progression of diseases.

- Identifying risk factors and biomarkers associated with certain conditions.
- Exploring the impact of interventions on patients' quality of life.

Introduction

Clinical research is a vital component of scientific exploration that significantly advances medical knowledge and enhances patient care. This field involves a wide range of studies aimed at assessing the safety, efficacy, and effectiveness of medical interventions, such as pharmaceuticals, therapies, medical devices, and behavioral treatments. Through meticulous data collection from human participants, clinical research provides essential evidence that informs medical decisions and improves patient outcomes.

Additionally, clinical research is fundamental in understanding disease mechanisms, identifying new therapeutic targets, and evaluating preventive measures. It includes various phases of clinical trials, from initial safety assessments in phase I to large-scale efficacy studies in phase III, and post-marketing surveillance in phase IV. This rigorous process ensures that new treatments are not only effective but also safe for public use. Furthermore, clinical research often involves interdisciplinary collaboration, incorporating insights from biostatistics, pharmacology, and ethics to uphold the integrity and applicability of its findings.

Understanding Clinical Research

At its core, clinical research entails the application of scientific methods to assess the impact and outcomes of medical interventions on human subjects. It strives to answer key questions related to the benefits, risks, mechanisms of action, and optimal use of these interventions in diverse patient populations. By employing well-designed protocols, robust data collection methods, and statistical analyses, clinical researchers seek to derive meaningful insights that can transform healthcare.

Objectives of Clinical Research

Safety Assessment

One of the primary goals of clinical research is to thoroughly evaluate the safety of medical interventions. This involves examining potential adverse effects, drug interactions, long-term consequences, and any other factors that may compromise patient well-being. By uncovering and understanding potential risks, clinical researchers help safeguard patient health and inform regulatory agencies.

Efficacy Evaluation

Clinical research investigates the efficacy of medical interventions by assessing their ability to produce desirable outcomes under controlled conditions. This includes determining whether a drug effectively treats a specific condition, a treatment modality leads to desired clinical responses, or a medical device performs as intended. Such evaluations provide clinicians and regulatory bodies with evidence-based information on which interventions are most effective.

Effectiveness Analysis

Apart from evaluating efficacy, clinical research also explores the effectiveness of interventions in real-world settings. This stage involves studying how medical interventions perform when applied to diverse patient populations, with varying demographics, comorbidities, and healthcare settings. Assessing effectiveness helps bridge the gap between controlled research environments and everyday clinical practice, ensuring that interventions work as expected in diverse scenarios.

Key Components of Clinical Research

Study Design

Clinical research encompasses various study designs, such as randomized controlled trials (RCTs), observational studies, case-control studies, cohort studies, and systematic reviews. Each design has specific strengths and limitations, and researchers select the most appropriate design based on the research question, available resources, and ethical considerations.

Informed Consent

Ethical considerations are of paramount importance in clinical research. Prior to participating in a study, individuals must provide informed consent, indicating their understanding of the study's purpose, procedures, potential risks, and benefits. This ensures respect for individual autonomy and protection of participants' rights.

Data Collection and Analysis

Rigorous data collection is crucial in clinical research to obtain accurate and reliable results. Researchers employ various methods, such as surveys, interviews, medical tests, and electronic health records, to gather relevant data. Statistical analysis techniques are then employed to interpret and draw meaningful conclusions from the collected data.

Conclusion

Clinical research serves as the foundation for evidence-based medicine, contributing to the advancement of medical knowledge and the improvement of patient care. By systematically investigating the safety, efficacy, and effectiveness of medical interventions, clinical research provides vital information to guide clinical decision-making, regulatory processes, and the development of new therapies. Its multifaceted nature, encompassing diverse study designs and ethical considerations, highlights the importance of robust methodologies and collaboration between researchers, healthcare professionals, and patients to ensure the highest standards of scientific inquiry.

SECTION 2: HISTORICAL MILESTONES IN CLINICAL RESEARCH

Historical Milestones in Clinical Research

In this section, we will explore the historical milestones that have shaped clinical research. We will discuss significant breakthroughs and achievements, such as the discovery of penicillin, the development of randomized controlled trials, and the establishment of ethical guidelines for human subjects' protection. By highlighting these milestones, we will emphasize the long-standing tradition of clinical research and its critical role in medical advancements.

Historical Milestones in Clinical Research: Penicillin, Randomized Controlled Trials, and Ethical Guidelines for Human Subjects' Protection.

Introduction

Clinical research has a rich history marked by several significant milestones that have shaped medical advancements and improved patient care. Three pivotal milestones that stand out are the discovery of penicillin, the development of randomized controlled trials (RCTs), the establishment of ethical guidelines for the protection of human subjects, advancements in genetic research, and the development of vaccines.

Discovery of Penicillin (1928)

One of the most remarkable breakthroughs in the history of clinical research was the discovery of penicillin by Sir Alexander Fleming in 1928. While conducting experiments on bacterial cultures, Fleming observed that a mold, later identified as Penicillium notatum, inhibited the growth of bacteria. This accidental discovery opened the door to the development of the first antibiotic.

Further research and clinical trials conducted by Howard Florey, Ernst Chain, and their team at the University of Oxford in the early 1940s demonstrated the therapeutic potential of penicillin. They successfully purified and produced penicillin in sufficient quantities to treat bacterial infections. This groundbreaking discovery revolutionized the treatment of infectious diseases, saving countless lives and establishing the era of antibiotics.

The discovery of penicillin exemplifies the vital role of clinical research in identifying new therapies, demonstrating their effectiveness, and transforming medical practice.

Development of Randomized Controlled Trials (RCTs)

Randomized controlled trials (RCTs) emerged as a key milestone in clinical research methodology during the mid-20th century. RCTs are experimental studies that involve randomly assigning participants to different treatment groups to assess the efficacy and safety of interventions.

The concept of randomization in clinical research can be traced back to the work of Sir Ronald Fisher in the 1920s. However, it was not until the 1940s that Austin Bradford Hill and Sir Richard Doll conducted a groundbreaking RCT to evaluate the effectiveness of streptomycin in treating tuberculosis. This study established the importance of randomization in minimizing bias and ensuring reliable results.

Since then, RCTs have become the gold standard for evaluating the effectiveness of medical interventions. They provide a robust framework for comparing treatment outcomes between groups, establishing causal relationships, and informing evidence-based medicine. RCTs have played a crucial role in the development of new drugs, treatments, and medical devices, leading to significant medical advancements.

The development of RCTs showcases how clinical research methodology has evolved and become a cornerstone of evidence-based medicine.

Establishment of Ethical Guidelines for Human Subjects' Protection

The protection of human subjects in clinical research became a significant concern following several unethical experiments in the mid-20th century, such as the Tuskegee Syphilis Study and the Nuremberg Trials. These egregious violations of ethical principles led to the establishment of guidelines and regulations to ensure the rights, safety, and well-being of participants in clinical research.

In 1964, the World Medical Association formulated the Declaration of Helsinki, a foundational document that outlined ethical principles for medical research involving human subjects. The declaration emphasizes the necessity of informed consent, the importance of minimizing risks, and the requirement for independent ethical review committees.

Subsequently, various countries and international bodies developed their own ethical guidelines. In the United States, the Belmont Report, published in 1979, outlined the fundamental ethical principles of respect for persons, beneficence, and justice.

These ethical guidelines served as a catalyst for change, promoting transparency, informed decision-making, and the establishment of institutional review boards (IRBs) or ethics committees. These bodies ensure that research studies meet ethical standards before initiation and provide ongoing oversight.

The establishment of ethical guidelines underscores the commitment of the scientific community to protect the rights and welfare of human subjects involved in clinical research. It reflects the recognition that ethical conduct is an essential prerequisite for generating reliable and trustworthy data.

Advancements in Genetic Research

The late 20th and early 21st centuries brought significant advancements in genetic research. The completion of the Human Genome Project in 2003 mapped the entire human genome, facilitating breakthroughs in understanding genetic predispositions to diseases, personalized medicine, and targeted therapies. This has transformed approaches to treatment and prevention in clinical practice.

Development of Vaccines

Clinical research has been instrumental in developing vaccines that have eradicated or controlled many infectious diseases. For instance, the development of the polio vaccine by Jonas Salk in the 1950s and subsequent vaccines has dramatically reduced the incidence of polio worldwide. More recently, the rapid development and testing of

COVID-19 vaccines demonstrated the power and importance of rigorous clinical research in responding to global health crises.

These historical milestones highlight the importance of clinical research in advancing medical science and improving patient outcomes. The ongoing commitment to ethical standards and methodological rigor ensures that clinical research continues to evolve, bringing new therapies and hope to patients worldwide. Additionally, clinical research has continuously pushed the boundaries of medical knowledge, introducing innovative therapies and improving patient outcomes. These milestones emphasize the significance of rigorous scientific inquiry, ethical considerations, and collaboration among researchers, healthcare professionals, and patients.

Conclusion

As we move forward, it is essential to build upon these milestones, promoting ongoing advancements in clinical research methodology, ethical standards, and the translation of research findings into tangible benefits for patients worldwide. The future of clinical research lies in embracing new technologies, personalized medicine, and patient-centered approaches, ensuring that research findings are effectively translated into improved healthcare practices and outcomes.

SECTION 3: THE SCIENTIFIC METHOD AND CLINICAL RESEARCH

The Scientific Method and Clinical Research

In this section, we will explain how clinical research follows the scientific method, a systematic approach to acquiring knowledge. We will discuss the key components of the scientific method, including:

- *Formulating a research question or hypothesis: We will explain how researchers formulate research questions based on existing knowledge gaps or hypotheses they seek to test.*
- *Study design and protocol development: We will discuss the importance of study design in clinical research, emphasizing the need for well-designed protocols that ensure reliable and valid results.*
- *Data collection and analysis: We will explain the process of gathering data through various methods, such as surveys, interviews, laboratory tests, and imaging techniques. Additionally, we will touch upon the statistical analysis used to interpret and draw conclusions from the collected data.*
- *Peer review and publication: We will discuss the critical role of*

peer review in evaluating the rigor and validity of research findings before publication in scientific journals.

- *Replication and validation: We will emphasize the importance of replication and independent validation of research findings to ensure the reliability and generalizability of results.*

Introduction

Clinical research follows the scientific method, a systematic approach to acquiring knowledge, to ensure the reliability, validity, and reproducibility of its findings. This methodical process involves several key steps that clinical researchers adhere to when designing, conducting, and interpreting their studies. Let's explore these steps in detail.

1. Observation and Question Formulation

The scientific method begins with observation, where researchers identify a phenomenon or problem of interest in the field of healthcare. This observation leads to the formulation of research questions or hypotheses. These questions guide the direction of the study and set the objectives to be addressed through the research process.

2. Literature Review

Before initiating a study, researchers conduct a comprehensive review of existing literature and research in the field. This step helps them understand the current state of knowledge, identify gaps, and build upon existing evidence. The literature review ensures that the research is informed by previous findings and contributes to the scientific knowledge base.

3. Study Design

Based on the research questions and available resources, researchers carefully design the study. The study design involves selecting appropriate methodologies, defining the study population, determining sample size, and outlining data collection methods. Common study designs in clinical research include randomized controlled trials, cohort studies, case-control studies, cross-sectional studies, and qualitative research methods, among others.

4. Data Collection

Data collection is a crucial phase of the scientific method in clinical research. Researchers use various tools, such as surveys, interviews, medical tests, and patient records, to collect relevant and reliable data. Data collection methods are chosen based on the study objectives, the nature of the research questions, and the available resources.

5. Data Analysis

After collecting the data, researchers employ statistical or qualitative analysis techniques to interpret and derive meaningful conclusions. Statistical analysis involves applying appropriate statistical tests to identify patterns, relationships, and statistical significance. Qualitative analysis focuses on understanding and interpreting non-numerical data, such as interviews or observations, to generate insights.

6. Results Interpretation and Conclusion

The interpretation of study results is a critical step in the scientific method. Researchers analyze and synthesize the findings, comparing them with existing literature and knowledge. They consider the limitations of the study and provide explanations for the observed outcomes. The conclusion of the study is based on the results and provides answers to the research questions or hypotheses.

7. Peer Review and Publication

To ensure the quality and credibility of research, the findings are typically subjected to peer review. Experts in the field critically evaluate the study design, methodology, analysis, and interpretation of the results. Feedback from the peer-review process helps researchers improve their work and address any weaknesses. Once accepted, the study is published in reputable scientific journals, contributing to the scientific knowledge base and allowing for replication and further scrutiny.

8. Replication and Meta-Analysis

Replication of studies by independent researchers is essential in the scientific method to validate and verify research findings. Replication helps establish the robustness and generalizability of results. Additionally, meta-analyses, which combine the results of multiple studies, provide a higher level of evidence and a more comprehensive understanding of the research question.

9. Iterative Process

The scientific method in clinical research is an iterative process, with each study building upon previous knowledge. New findings often lead to the formulation of new research questions, which in turn generate new studies. This iterative process drives the advancement of medical knowledge and the continuous improvement of patient care.

Conclusion

By following the scientific method, clinical research ensures that investigations are conducted systematically, promoting rigorous methods, reproducibility, and the accumulation of evidence-based knowledge. This methodical approach contributes to the credibility and reliability of research findings, leading to improvements in medical interventions and patient outcomes.

SECTION 4: KEY TERMINOLOGIES IN CLINICAL RESEARCH

Key Terminologies in Clinical Research

This section will familiarize readers with essential terminologies commonly used in clinical research. We will define and explain terms such as Protocol, Informed Consent, Placebo, Blinding, Randomization, Sample Size, and Endpoint.

Protocol

A research protocol is a comprehensive document that outlines the plan and details of a research study. It serves as a roadmap for conducting the study, providing a clear and structured framework for researchers to follow. The protocol includes essential information about the study objectives, design, procedures, and participant eligibility criteria. Let's explore the purpose and key components of a research protocol in detail.

1. Study Objectives

The protocol clearly states the primary and secondary objectives of the study. These objectives define the specific research questions that the study aims to answer or the hypotheses it seeks to test. The objectives provide a clear focus for the study and guide the design and execution of the research.

2. Study Design

The protocol describes the study design, which is the overall plan for how the study will be conducted. Common study designs include randomized controlled trials, observational studies, cohort studies, case-control studies, and cross-sectional studies. The design details the methods for participant allocation, data collection, and data analysis, ensuring that the study design aligns with the research objectives.

3. Procedures and Methods

The protocol outlines the procedures and methods that will be used to collect and analyze data. It specifies the data collection instruments, such as questionnaires, surveys, medical tests, or observations, and explains how the data will be recorded and managed. The protocol also describes any interventions, treatments, or experimental procedures that will be administered to participants.

4. Participant Eligibility Criteria

The protocol defines the criteria that participants must meet to be eligible for the study. These criteria may include demographic characteristics, medical history, specific clinical conditions, or other relevant factors. Participant eligibility criteria ensure that the study population is appropriate for addressing the research objectives and that the findings can be generalized to the intended population.

5. Ethical Considerations

The protocol addresses ethical considerations and ensures that the study adheres to ethical guidelines and regulations. It explains how informed consent will be obtained from participants, how confidentiality and privacy will be protected, and how potential risks and benefits will be communicated. Ethical considerations are crucial for protecting the rights and welfare of research participants.

6. Data Analysis Plan

The protocol outlines the planned statistical or qualitative analysis methods that will be used to analyze the collected data. It specifies the statistical tests, measures of association, or qualitative analysis techniques that will be employed to answer the research questions or test the hypotheses. The data analysis plan ensures that the collected data will be appropriately analyzed to derive meaningful conclusions.

7. Timeline and Resources

The protocol includes a timeline outlining the anticipated duration of the study, from participant recruitment to data analysis and reporting. It also specifies the resources required, such as funding, personnel, equipment, and facilities, to conduct the study. This information helps in planning and budgeting for the study and ensures that resources are allocated appropriately.

8. Quality Control and Monitoring

The protocol describes strategies for quality control and study monitoring. It outlines procedures for data verification, data management, and quality assurance. It may also include plans for regular monitoring visits, audits, or interim analyses to assess the progress of the study and ensure data integrity.

9. Dissemination of Findings

The protocol may outline the planned dissemination of study findings, including publication in scientific journals, presentations at conferences, or reports to relevant stakeholders. Dissemination of findings is essential for sharing knowledge and contributing to the scientific community.

In summary, a research protocol plays a crucial role in guiding the planning, execution, and reporting of a research study. It ensures that the study objectives, design, procedures, and participant eligibility criteria are clearly defined and aligned with ethical considerations. By providing a structured framework, the protocol facilitates the successful implementation of the study and helps ensure the validity and reliability of the research findings.

Informed Consent

Informed consent is a fundamental ethical principle in research involving human subjects. It is a process through which individuals voluntarily and comprehensively provide their explicit agreement to participate in a research study. Informed consent ensures that participants understand the nature of the study, its objectives, procedures, potential risks and benefits, and their rights as research subjects. Let's explore the significance and key components of informed consent in detail:

Voluntary Participation

Informed consent emphasizes the voluntary nature of participation in research. It ensures that individuals have the freedom to decide whether or not to participate, without any coercion or undue influence. Participants should not feel pressured or compelled to join a study but rather make an autonomous decision based on their own judgment.

Understanding the Nature of the Study

Informed consent requires researchers to provide clear and accessible information about the study. This includes explaining the purpose of the research, the study procedures, the expected duration, and any interventions or treatments involved. Participants should have a comprehensive understanding of what their involvement entails.

Risks and Benefits

Participants must be informed about the potential risks and benefits associated with their participation in the study. Researchers should clearly communicate any known or foreseeable risks, side effects, discomfort, or inconvenience that participants may experience. Additionally, participants should be aware of any potential benefits, such as access to new treatments or contributing to scientific knowledge.

Confidentiality and Privacy

Informed consent addresses the protection of participants' confidentiality and privacy. Participants should be informed about how their personal information will be handled, stored, and anonymized to ensure confidentiality. They should also understand any limitations to confidentiality, such as instances where disclosure may be required by law or when there is a concern for their safety.

Rights and Protections

Informed consent highlights the rights and protections afforded to research participants. This includes the right to withdraw from the study at any time without penalty or repercussion. Participants should be aware of their right to ask questions, seek clarifications, and request additional information throughout the study. They should also understand that their decision to participate or not will not affect their regular medical care or their relationship with healthcare providers.

Comprehension and Communication

Informed consent requires that researchers use clear, understandable, and non-technical language when explaining the study to participants. Researchers should make an effort to ensure that participants comprehend the information provided. This may involve using visual aids, interpreters, or simplifying complex concepts to enhance understanding.

Documentation

Informed consent is typically documented through a written consent form signed by the participant or their authorized representative. This form serves as evidence that the participant has been adequately informed and has voluntarily consented to participate. The consent form should accurately reflect the information provided during the consent process and should be easily accessible for participants to refer to throughout the study.

The significance of informed consent in research cannot be overstated. It upholds ethical principles of respect for autonomy, dignity, and human rights. Informed consent empowers participants by ensuring that they have the necessary information to make informed decisions about their involvement in research. It also promotes transparency, trust, and mutual respect between researchers and participants, fostering a collaborative and ethical research environment.

Placebo

In clinical research, a placebo is an inactive substance or intervention that has no therapeutic effect on the condition being studied. It is used as a control in clinical trials to assess the efficacy of an active intervention, such as a drug or treatment. The concept of placebos plays a crucial role in the design and interpretation of clinical research studies. Let's explore the concept of placebos in clinical research in more detail:

Control Group

In clinical trials, participants are often divided into two groups: the treatment group and the control group. The treatment group receives the active intervention being studied, while the control group receives a placebo. The purpose of using a placebo is to establish a baseline against which the effects of the active intervention can be compared. By comparing the outcomes between the treatment group and the control group, researchers can determine the true effects of the intervention.

Inert Substances or Interventions

Placebos can take various forms, depending on the nature of the study. In drug trials, a placebo may be an inert substance, such as a sugar pill, saline solution, or an inactive filler. In medical device trials, a placebo may involve using a device that appears identical to the active device but lacks the active components or functions. Placebos can also be sham procedures or simulated interventions that mimic the active intervention but have no physiological effect.

Blinding

To ensure unbiased results, clinical trials often employ a technique called blinding or masking. Blinding involves withholding information about whether participants are receiving the active intervention or the placebo. In a single-blind study, participants are unaware of their group assignment. In a double-blind study, both participants and researchers are unaware of who is receiving the active intervention or the placebo. Blinding helps minimize biases and ensures that the observed effects are not influenced by the participants' or researchers' expectations.

Placebo Effect

The placebo effect refers to the phenomenon where a placebo, despite having no therapeutic effect, can produce a perceived improvement in a patient's symptoms. This effect is believed to result from the patient's expectation of improvement, the psychological influence of participating in a study, or other non-specific factors. The placebo effect highlights the importance of using a control group in clinical research to distinguish the true effects of an intervention from the effects attributed to placebo responses.

Ethical Considerations

The use of placebos in clinical research raises ethical considerations. In some cases, it may be ethically inappropriate to use a placebo, particularly when an effective treatment already exists for a condition. In such instances, it is essential to consider the ethical principle of equipoise, which requires that the equipoise or uncertainty exists about the comparative effectiveness of the interventions being studied. Ethical guidelines and review boards play a crucial role in evaluating the appropriateness of using placebos and ensuring participant safety and welfare.

Advancements in Study Design

In recent years, there has been an increasing emphasis on the use of active comparators in clinical trials rather than relying solely on placebo controls. Active comparators involve comparing the effectiveness of a new intervention against an existing standard treatment or an established therapy. This approach ensures that participants receive an active intervention and avoids withholding potentially beneficial treatments from participants.

In conclusion, placebos are inert substances or interventions used as controls in clinical research to assess the efficacy of active interventions. They serve as a baseline against which the effects of the active intervention are evaluated. Placebos are employed in clinical trials to minimize biases, evaluate the true effects of interventions, and distinguish them from placebo effects. Ethical considerations and advancements in study design continue to shape the use of placebos in clinical research.

Blinding

Blinding, also known as masking, is a technique used in clinical trials to minimize bias and ensure the integrity of study results. It involves withholding information about the assigned interventions from participants and/or researchers. Blinding helps prevent conscious or subconscious biases that can influence participant behavior or researcher assessments. There are two main types of blinding: single-blind and double-blind studies.

Single-blind Studies

In a single-blind study, participants are unaware of their assigned interventions, while researchers are aware of the assignments. This means that participants do not know whether they are receiving the active treatment or a placebo/control. The purpose of single-blinding is to prevent participants' knowledge of their treatment from influencing their behaviors, expectations, or reporting of outcomes. By keeping participants blind to their interventions, researchers can minimize placebo effects or other biases related to participant awareness.

For example, in a drug trial, participants in the treatment group might receive the active medication, while those in the control group receive a placebo. However, neither group knows which treatment they are receiving. This blinding helps ensure that any observed differences in outcomes between the groups are not influenced by participants' beliefs or expectations.

Double-blind Studies

In a double-blind study, both participants and researchers are unaware of the assigned interventions. This means that neither the participants nor the investigators know whether the participants are receiving the active treatment or the placebo/control. Double-blinding aims to prevent biases that may arise from participants' expectations or researchers' assessments.

To achieve double-blinding, a third party, such as a pharmacist or an independent statistician, is often responsible for the randomization and assignment of participants to interventions. They maintain the key to the treatment assignments and ensure that the study participants and researchers remain unaware of the allocations throughout the study duration.

Double-blinding is considered the gold standard for clinical trials because it helps reduce various sources of bias, such as the placebo effect, observer bias, and confirmation bias. It enhances the scientific rigor and validity of the study results by ensuring that the observed outcomes are truly attributed to the interventions being tested.

Blinding in both single-blind and double-blind studies helps maintain the integrity and credibility of research findings. By minimizing biases related to participant and researcher awareness, blinding ensures that the observed effects are more likely to be a result of the interventions themselves rather than influenced by external factors. It allows for a fair and unbiased evaluation of the interventions' efficacy or effectiveness.

It's important to note that blinding is not always feasible or appropriate, particularly in certain surgical procedures or behavioral interventions where the nature of the intervention makes it challenging to blind participants or researchers. In such cases, researchers employ other methods to minimize bias, such as standardized protocols, independent outcome assessments, or the use of objective endpoints.

Randomization

Randomization is a fundamental principle in clinical trials that helps ensure unbiased and reliable study results. It involves the random assignment of participants to different treatment groups or study arms. The process of randomization is designed to minimize the influence of confounding factors and potential biases, enhancing the validity and internal validity of the study. Let's explore the concept of randomization in detail:

Purpose of Randomization

The primary purpose of randomization is to create treatment groups that are comparable in terms of baseline characteristics and potential prognostic factors. Random assignment helps distribute both known and unknown factors equally among the treatment groups, reducing the potential for systematic differences that could influence study outcomes. By distributing these factors randomly, researchers can attribute any differences observed between the groups to the interventions being studied rather than to other factors.

Unbiased and Unpredictable Allocation

Randomization ensures that the assignment of participants to treatment groups is unbiased and unpredictable. It eliminates selection bias, where researchers or participants may consciously or unconsciously assign participants to a particular group based on their characteristics, preferences, or expectations. Random allocation ensures that each participant has an equal chance of being assigned to any of the treatment groups, thereby eliminating systematic differences between the groups.

Randomization Methods

Randomization can be achieved using various methods, depending on the specific study design and practical considerations. The commonly used randomization methods include:

Simple Randomization

In simple randomization, participants are assigned to treatment groups by a process similar to flipping a coin or drawing lots. This method ensures each participant has an equal and independent chance of being assigned to any treatment group.

Stratified Randomization

Stratified randomization is used when certain participant characteristics are known to significantly affect the outcomes. Participants are first grouped into strata based on these characteristics (e.g., age, gender, disease severity), and then randomization is performed within each stratum. This ensures that each treatment group has a balanced distribution of participants with similar characteristics.

Blocked Randomization

Blocked randomization is used to ensure balance and equal group sizes as participants are enrolled in the study. It involves creating blocks of participants (e.g., blocks of four or six) and randomly assigning them to treatment groups within each block. This method helps maintain approximate equal group sizes throughout the study.

Adaptive Randomization

Adaptive randomization is a flexible method that adjusts the allocation probabilities based on accumulated data during the study. It allows for a higher probability of assigning participants to the treatment group that appears to be more effective or beneficial based on interim analysis. Adaptive randomization is commonly used in adaptive clinical trials and can maximize the efficiency of the study.

Allocation Concealment

To maintain the integrity of randomization, allocation concealment is crucial. It refers to the process of keeping the randomization scheme hidden from researchers, participants, and other study personnel involved in participant enrollment. This prevents selection bias or deliberate manipulation of group assignments. Techniques such as central randomization systems, sealed envelopes, or online randomization services can be employed to ensure proper allocation concealment.

Blinding and Randomization

Randomization is often combined with blinding to further enhance the validity of study results. Blinding ensures that both participants and researchers are unaware of the assigned interventions, minimizing biases related to expectations or assessments. Blinding can be achieved through single-blind or double-blind study designs, as discussed in a previous response.

By employing randomization in clinical trials, researchers can obtain reliable and unbiased estimates of treatment effects. Random allocation helps create comparable treatment groups, minimizing confounding factors and enhancing internal validity. It allows for a rigorous evaluation of the interventions being studied and provides a solid foundation for drawing valid conclusions about their efficacy or effectiveness.

Sample Size

Sample size refers to the number of participants or study units included in a clinical trial or research study. Determining an appropriate sample size is a critical aspect of study design, as it directly impacts the statistical power and precision of the study's findings. Let's explore the concept of sample size in detail and understand its importance:

Statistical Power

Statistical power refers to the ability of a study to detect a true effect or difference between the treatment groups when it exists. Adequate statistical power is essential to draw reliable conclusions from a study. Insufficient sample size can lead to low statistical power, increasing the likelihood of false-negative results (i.e., failure to detect a real effect). By determining an appropriate sample size, researchers can enhance the statistical power of their study, ensuring that they have a higher chance of detecting meaningful differences if they exist.

Precision of Estimates

Sample size also affects the precision of the estimates derived from a study. Precision refers to the degree of uncertainty or variability associated with the study findings. A larger sample size generally leads to narrower confidence intervals around the estimated treatment effects. Narrow confidence intervals indicate a higher level of precision, providing more precise and reliable estimates of the true treatment effects or outcomes. Conversely, smaller sample sizes may result in wider confidence intervals and increased uncertainty in the study findings.

Effect Size and Variability

Determining the appropriate sample size requires consideration of the anticipated effect size and the variability of the outcomes being measured. Effect size refers to the magnitude of the treatment effect or the difference between the groups being compared. Larger effect sizes generally require smaller sample sizes to detect the effects with sufficient power. Conversely, smaller effect sizes may necessitate larger sample sizes to achieve adequate power.

Variability, or the spread of data, within the sample also influences the required sample size. Greater variability generally requires larger sample sizes to achieve adequate power, as a larger sample is needed to detect smaller differences within the data.

Statistical Tests and Significance Level

The choice of statistical tests and the desired level of statistical significance also influence the determination of sample size. Different statistical tests have varying requirements for sample size calculations. Additionally, the desired level of significance (e.g., p-value threshold) affects the sample size determination. A lower significance level (e.g., $p<0.01$) may require a larger sample size compared to a higher significance level (e.g., $p<0.05$) to achieve the desired power.

Type I and Type II Errors

The determination of sample size is also influenced by the acceptable levels of Type I and Type II errors. Type I error occurs when a true null hypothesis is incorrectly rejected (false-positive result), while Type II error occurs when a false null hypothesis is not rejected (false-negative result). By calculating the sample size based on the desired power and the acceptable levels of Type I and Type II errors, researchers can strike an appropriate balance between these two types of errors.

The study design, including the recruitment strategy, resources, and ethical considerations, also plays a role in determining the sample size. Researchers need to consider the feasibility of enrolling and retaining the required number of participants within the available timeframe and resources. Practical considerations, such as the costs and logistics associated with conducting the study, should be considered when determining the sample size.

In summary, determining an appropriate sample size is crucial in clinical trials and research studies. It ensures adequate statistical power to detect true effects, enhances the precision of estimates, and minimizes the likelihood of false-negative or false-positive results. The sample size calculation should consider factors such as the anticipated effect size, variability, statistical tests, significance level, and the acceptable levels of Type I and Type II errors. Balancing these considerations leads to a sample size that optimizes the study's validity, reliability, and generalizability of the findings.

Endpoint

In a clinical trial, an endpoint is a specific event, outcome, or measurement used to evaluate the effectiveness or safety of an intervention being studied. Endpoints serve as objective measures that help researchers assess the impact of the intervention on the study participants. Let's explore the concept of endpoints in detail:

Clinical Endpoints

Clinical endpoints are outcome measures that directly reflect the impact of the intervention on the participants' health, well-being, or disease status. These endpoints are typically patient-centered and relevant to the condition being treated. Clinical endpoints provide information on the efficacy, safety, or both of the intervention under investigation. They are often the primary focus of the study and serve as the basis for determining the success or failure of the intervention.

Clinical endpoints can take various forms depending on the nature of the study and the condition being evaluated. Some examples of clinical endpoints include:

1. **Survival:** The time to an event, such as death or disease recurrence, is a common clinical endpoint in studies involving life-threatening conditions.
2. **Disease Progression:** Clinical endpoints can measure the progression of a disease, such as the increase in tumor size or the worsening of symptoms.
3. **Symptom Improvement:** Endpoints that assess the relief of symptoms, such as pain reduction, improvement in quality of life, or decreased frequency of symptoms, are important in many clinical trials.
4. **Functional Outcomes:** These endpoints assess improvements in functional abilities, such as mobility, cognitive function, or physical performance.

5. **Laboratory Measurements:** Certain clinical endpoints involve specific laboratory measurements, such as cholesterol levels, blood pressure, or biomarker concentrations.

Primary Endpoint

The primary endpoint in a clinical trial is the main outcome measure that is of greatest importance in determining the efficacy or effectiveness of the intervention. It is selected based on the research question, clinical relevance, and statistical considerations. The primary endpoint is typically the one used for sample size calculations and determining the success criteria of the study. The results of the primary endpoint are considered pivotal in making conclusions about the intervention's efficacy or effectiveness.

For example, in a study evaluating a new drug for hypertension, the primary endpoint might be the reduction in systolic blood pressure after a specified duration of treatment. The study's success or failure would be determined based on the achievement of a statistically significant reduction in blood pressure as the primary endpoint.

Secondary Endpoint

Secondary endpoints are additional outcome measures that provide supportive or complementary information to the primary endpoint. They offer a more comprehensive evaluation of the intervention's effects on various aspects of the disease or patient outcomes. Secondary endpoints may provide insights into specific subgroups of patients, alternative treatment effects, or exploratory findings.

Secondary endpoints are typically pre-specified but are considered of lesser importance than the primary endpoint. They can help researchers gather additional data, generate hypotheses, or explore potential benefits or risks associated with the intervention. However, it's important to note that statistical analyses of secondary endpoints may require adjustments to account for multiple comparisons.

For example, in a clinical trial for a new cancer treatment, the primary endpoint might be overall survival, while secondary endpoints could include progression-free survival, tumor response rates, or quality of life assessments.

Composite Endpoints

In some studies, composite endpoints are used to capture multiple outcomes or events within a single measure. Composite endpoints are useful when individual outcomes may be infrequent, require long follow-up periods, or are not sufficiently powered to detect differences. These endpoints combine multiple related events or outcomes into a single measure, providing a more comprehensive assessment of the intervention's effects.

For instance, in cardiovascular trials, a composite endpoint may include a combination of outcomes such as cardiovascular death, non-fatal heart attack, stroke, or hospitalization for heart failure. The composite endpoint offers a more comprehensive evaluation of the treatment's impact on cardiovascular events.

In conclusion, endpoints in clinical trials are specific measures used to evaluate the outcomes of the study. Clinical endpoints directly reflect the impact of the intervention on participants' health or disease status. Primary endpoints are the main outcome measures used to determine the intervention's efficacy or effectiveness, while secondary endpoints provide additional supportive or exploratory information. Composite endpoints combine multiple outcomes into a single measure, offering a comprehensive evaluation of the intervention's effects.

SECTION 5: ETHICAL CONSIDERATIONS IN CLINICAL RESEARCH

Ethical Considerations in Clinical Research

In this section, we will delve into the ethical considerations that underpin clinical research. We will discuss the historical context that led to the development of ethical guidelines, such as the Nuremberg Code and the Declaration of Helsinki. We will explore key ethical principles, including respect for autonomy, beneficence, and justice, and explain how these principles guide the conduct of research involving human participants. We will emphasize the importance of protecting participants' rights, ensuring informed consent, maintaining confidentiality, and minimizing risks while maximizing benefits.

Ethical Considerations

Ethical considerations play a crucial role in clinical research to ensure the protection of human subjects and the integrity of scientific inquiry. Over the years, several historical events have shaped the development of ethical guidelines, leading to the establishment of key documents such as the Nuremberg Code and the Declaration of

Helsinki. Let's delve into the historical context and ethical considerations that underpin clinical research:

Nuremberg Code

The Nuremberg Code, developed in response to the atrocities committed during World War II, laid the foundation for ethical principles in human experimentation. The Nuremberg Trials held from 1945 to 1946 brought to light the heinous experiments conducted by Nazi doctors on prisoners without their informed consent.

The Nuremberg Code, published in 1947, established ten principles that should guide human experimentation. These principles included voluntary informed consent, the importance of scientific justification and expertise, minimizing risks, and ensuring the well-being of the participants. The code emphasized that experiments should be based on sound scientific principles and should not cause unnecessary harm or suffering.

Declaration of Helsinki

The Declaration of Helsinki, first adopted in 1964 and subsequently revised, is a cornerstone document in medical research ethics. It was developed by the World Medical Association (WMA) as a set of guidelines to protect the rights and welfare of research participants.

The Declaration of Helsinki addresses key ethical considerations in clinical research, including informed consent, beneficence, respect for the individual, and scientific validity. It emphasizes the importance of balancing the benefits and risks of research, ensuring that research is based on prior animal and laboratory experimentation, and that the risks are justified by potential benefits to society. The Declaration of Helsinki also introduced the concept of an independent ethical review committee to assess research protocols and ensure compliance with ethical principles.

Belmont Report

In the United States, the Belmont Report, published in 1979, further solidified ethical principles in research involving human subjects. The report was a response to the unethical experiments such as the Tuskegee Syphilis Study and other controversial studies.

The Belmont Report established three fundamental ethical principles: respect for persons, beneficence, and justice. Respect for persons emphasizes the autonomy and informed consent of participants. Beneficence focuses on maximizing benefits and minimizing risks, ensuring the well-being of participants. Justice addresses the fair distribution of the benefits and burdens of research.

Institutional Review Boards (IRBs)

To implement and enforce ethical guidelines, Institutional Review Boards (IRBs) were established. IRBs are independent committees responsible for reviewing research protocols, ensuring that studies meet ethical standards, and safeguarding the rights and welfare of participants. IRBs assess the informed consent process, risks and benefits, study design, and ongoing monitoring of research.

International Ethical Guidelines

In addition to the Nuremberg Code and the Declaration of Helsinki, several international organizations and regulatory bodies have developed ethical guidelines for clinical research. These include the International Council for Harmonization of Technical Requirements for Pharmaceuticals for Human Use (ICH), the World Health Organization (WHO), and the Council for International Organizations of Medical Sciences (CIOMS).

These guidelines emphasize the importance of protecting vulnerable populations, ensuring informed consent, promoting transparency, and maintaining confidentiality. They provide ethical frameworks for research involving different populations, such as children, pregnant women, and individuals with diminished decision-making capacity.

Ongoing Ethical Considerations

Ethical considerations in clinical research continue to evolve in response to emerging challenges, technological advancements, and changing societal norms. Contemporary discussions include issues related to privacy and data protection, access to experimental treatments, genetic research, and the involvement of marginalized populations.

Efforts are ongoing to ensure that ethical principles are upheld in clinical research globally. Collaboration among researchers, regulatory bodies, and ethical review committees is vital to promote ethical conduct, protect participants, and advance scientific knowledge while upholding the principles of autonomy, beneficence, and justice.

Conclusion

In summary, ethical considerations in clinical research have a rich historical context that emerged from tragic events and unethical practices. The Nuremberg Code, Declaration of Helsinki, Belmont Report, and international guidelines provide a framework for protecting human subjects and ensuring ethical conduct in research. These guidelines underscore the importance of informed consent, scientific validity, beneficence, and justice, and continue to evolve in response to the changing landscape of medical research.

Ethical Principles

Ethical principles serve as a guiding framework for the conduct of research involving human participants. Three key ethical principles that underpin clinical research are respect for autonomy, beneficence, and justice. Let's explore each of these principles in detail and understand how they inform the ethical conduct of research.

Respect for Autonomy

Respect for autonomy recognizes the individual's right to self-determination, freedom of choice, and respect for their decision-making capacity. It acknowledges that individuals have the right to make informed decisions about their participation in research without coercion or undue influence.

In the context of research, respect for autonomy involves:

1. **Informed Consent:** Researchers must provide participants with comprehensive information about the study, including its purpose, procedures, risks, benefits, and alternatives. Participants should have the capacity to understand the information and voluntarily provide their informed consent to participate or withdraw from the study at any time.
2. **Privacy and Confidentiality:** Researchers must protect the privacy and confidentiality of participants' personal information. They should ensure that participant data is securely stored, anonymized when appropriate, and used solely for research purposes.
3. **Respect for Vulnerable Populations:** Special considerations are necessary when involving vulnerable populations, such as children, individuals with cognitive impairments, or those with limited decision-making capacity. Additional safeguards should be in place to

protect their rights and well-being, including obtaining proxy consent from legally authorized representatives.

Beneficence

Beneficence entails acting in the best interests of participants, promoting their well-being, and maximizing benefits while minimizing risks. It involves an obligation to do good and ensure that the potential benefits of the research outweigh the potential harms.

In the context of research, beneficence involves:

1. **Risk-Benefit Assessment:** Researchers should carefully assess the potential risks and benefits of the study to ensure that the anticipated benefits justify the potential harm or discomfort to participants. They should design studies with an aim to minimize risks and maximize potential benefits.
2. **Monitoring Participant Safety:** Researchers have an ongoing responsibility to monitor participant safety and well-being throughout the study. Appropriate measures should be in place to identify and address any adverse events or unanticipated risks promptly.
3. **Data Integrity and Quality:** Beneficence also includes conducting research with scientific rigor and integrity to generate reliable and valid results. Researchers should adhere to robust study design, appropriate data collection methods, and rigorous analysis to ensure the accuracy and credibility of findings.

Justice

The principle of justice emphasizes fairness in the distribution of benefits and burdens of research. It requires that research participants be selected in an equitable manner and that the benefits derived from research be shared fairly.

In the context of research, justice involves:

1. **Selection of Participants:** Researchers should ensure that the selection of participants is fair and does not exploit or discriminate against any particular population. Inclusion criteria should be based on scientific justification and relevant considerations rather than discriminatory factors.
2. **Access to Benefits:** Researchers should strive to ensure that the benefits resulting from research are accessible to all relevant populations, including those who participated in the study. This includes equitable access to any successful interventions or treatments that may emerge from the research.
3. **Fair Allocation of Resources:** Justice also encompasses the fair allocation of resources for research. Researchers should consider the allocation of resources in a way that minimizes disparities and ensures equitable access to research opportunities across different populations.
4. **Community Engagement:** Justice extends beyond individual participants to the broader community. Researchers should engage with the community and relevant stakeholders to understand their needs, concerns, and priorities. This engagement fosters collaboration, transparency, and accountability in the research process.

Conclusion

By adhering to the principles of respect for autonomy, beneficence, and justice, researchers can ensure the ethical conduct of research involving human participants. These principles provide a moral compass, guiding researchers to protect the rights, welfare, and dignity of participants while advancing scientific knowledge for the benefit of society. Ethical review boards and regulatory bodies play a crucial role in ensuring compliance with these principles and promoting ethical conduct in research.

Additional Ethical Considerations

Protecting Participants' Rights

Protecting participants' rights is of utmost importance in clinical research. Researchers must recognize and respect the inherent dignity, autonomy, and welfare of each participant. This includes ensuring that participants are treated with respect, their privacy is maintained, and their well-being is prioritized throughout the research process. Participants have the right to withdraw from the study at any time without facing repercussions.

Ensuring Informed Consent

Obtaining informed consent is a fundamental ethical requirement in research involving human participants. Informed consent ensures that participants are fully informed about the study's purpose, procedures, potential risks and benefits, and their rights as research subjects. Researchers should provide information in a clear and understandable manner, allowing participants to make voluntary decisions about their participation. Informed consent is an ongoing process, and participants should be given the opportunity to ask questions and seek clarification throughout the study.

Maintaining Confidentiality

Maintaining the confidentiality of participants' personal information is crucial to protect their privacy and uphold their rights. Researchers must ensure that participant data is securely stored, only accessible to authorized personnel, and used strictly for research purposes. Confidentiality extends beyond the research team to any publications or presentations of research findings, where participant identities should be protected through anonymization or pseudonymization.

Minimizing Risks while Maximizing Benefits

Ethical research requires a careful balance between the potential risks and benefits to participants. Researchers should make every effort to minimize risks by employing appropriate study design, considering the welfare of participants, and implementing safeguards to mitigate potential harm. This includes regularly monitoring participant safety, promptly addressing adverse events, and providing necessary medical care when required. Researchers should also strive to maximize the benefits of research by conducting studies that have the potential to contribute to scientific knowledge, improve medical treatments, or address significant health issues.

Incorporating Ethical Considerations into Study Design

Ethical considerations should be integrated into the study design from the outset. Researchers should carefully evaluate the study's objectives, methodology, and procedures to ensure they align with ethical principles. This includes considering the potential impact on participants, selecting appropriate control groups, and determining the sample size that provides valid results while minimizing participant burden.

Oversight and Ethical Review

To ensure adherence to ethical principles, research studies involving human participants are subject to oversight by independent ethical review boards or institutional review boards (IRBs). These boards review research protocols to evaluate their scientific and ethical rigor, assess the informed consent process, and monitor participant rights and welfare throughout the study. Ethical review boards play a crucial role in safeguarding participants' rights and ensuring that research meets the highest ethical standards.

By upholding participants' rights, obtaining informed consent, maintaining confidentiality, and carefully balancing risks and benefits, researchers can ensure that the well-being and dignity of participants are protected in clinical research. These ethical considerations not only ensure compliance with ethical guidelines but also contribute to the overall validity, integrity, and credibility of research outcomes.

SECTION 6: CHALLENGES AND CONTROVERSIES IN CLINICAL RESEARCH

Challenges and Controversies in Clinical Research

This section will shed light on the challenges and controversies that researchers and participants may encounter in the realm of clinical research. We will discuss issues such as participant recruitment, access to experimental treatments, conflicts of interest, publication bias, and the balance between scientific progress and participant welfare. By addressing these challenges, we aim to provide readers with a realistic understanding of the complexities inherent in clinical research.

Introduction

Clinical research is a complex field that presents various challenges and controversies for both researchers and participants. By shedding light on these issues, we can gain a realistic understanding of the complexities inherent in clinical research. Let's explore some of the key challenges and controversies in the realm of clinical research.

Participant Recruitment

Participant recruitment is a critical aspect of clinical research, but it can be challenging to identify and enroll a diverse and representative sample. Recruitment difficulties can arise due to strict inclusion and exclusion criteria, reluctance or skepticism among potential participants, limited access to certain populations, or a lack of awareness about research opportunities. Inadequate participant recruitment may compromise the generalizability and external validity of study findings.

Access to Experimental Treatments

In some cases, clinical trials involve experimental interventions that may hold promise for participants with serious or life-threatening conditions. However, limited access to these experimental treatments raises ethical and practical concerns. Issues such as eligibility criteria, availability of treatment slots, financial barriers, and geographical disparities can create challenges in ensuring equitable access to experimental interventions, potentially leaving some individuals without viable treatment options.

Conflicts of Interest

Conflicts of interest can arise when researchers or institutions have financial or other competing interests that may unduly influence the design, conduct, or reporting of clinical research. Financial ties to pharmaceutical companies, for example, can introduce biases and raise questions about the objectivity and integrity of the research. Managing conflicts of interest and promoting transparency are essential for maintaining the trust and credibility of clinical research.

Publication Bias

Publication bias refers to the selective publication of research findings based on their direction or significance, leading to an overrepresentation of positive or statistically significant results in the literature. This bias can distort the overall evidence base, as negative or non-significant findings may go unpublished or remain inaccessible. Publication bias can arise due to various factors, including publication preferences of journals, researcher biases, and industry influences. It hampers the ability to make informed decisions based on the totality of the evidence.

Balance between Scientific Progress and Participant Welfare

Clinical research aims to advance scientific knowledge and improve patient care, but striking a balance between scientific progress and participant welfare is crucial. Researchers must ensure that the risks to participants are minimized, benefits are maximized, and the study design is ethically sound. Ethical dilemmas may arise when determining the appropriate level of risk for participants, especially in early-phase trials or studies involving vulnerable populations. Ethical oversight and robust ethical review processes help navigate these challenges.

Informed Consent Challenges

Obtaining informed consent is essential to protect participants' rights and ensure their voluntary participation in research. However, challenges can arise in the informed consent process. Participants may face difficulties understanding complex study information, potential risks, and benefits. Language barriers, low health literacy, or power dynamics between researchers and participants can further complicate the process. Researchers must take extra care to ensure that informed consent is truly informed and voluntarily given.

Ethical Considerations for Placebo-Controlled Trials

In placebo-controlled trials, participants may be assigned to a control group receiving a placebo rather than an active intervention. This raises ethical concerns, particularly when effective treatments already exist for the condition being studied. Balancing the potential benefits of generating robust evidence with the ethical obligation to provide the best available care to participants can be challenging. Alternative study designs, such as comparative effectiveness research, may be considered to address these concerns.

Data Sharing and Transparency

Data sharing and transparency are essential for scientific progress, replication of findings, and avoiding unnecessary duplication of research. However, challenges related to data ownership, privacy concerns, intellectual property rights, and academic competition can hinder the sharing of research data. Striking a balance between protecting participants' privacy and promoting data sharing for the greater scientific good remains an ongoing challenge.

Long-Term Follow-up and Post-Market Surveillance

Long-term follow-up and post-market surveillance are critical for assessing the long-term safety and effectiveness of interventions beyond the duration of clinical trials. Challenges include participant retention, funding constraints, and ensuring the sustainability of long-term studies. Monitoring the real-world outcomes of interventions and identifying potential long-term risks or benefits require robust systems and collaboration among researchers, regulatory agencies, and healthcare providers.

Wrapping Up

Addressing these challenges and controversies in clinical research requires ongoing dialogue, collaboration, and regulatory oversight. Efforts to improve participant recruitment strategies, promote transparency, manage conflicts of interest, address publication bias, and strengthen ethical review processes are crucial for enhancing the quality, integrity, and ethical conduct of clinical research. Researchers, regulatory bodies, research institutions, and the scientific community as a whole must work together to navigate these complexities and uphold the highest ethical standards in clinical research.

By the end of Chapter 1, readers will have a solid foundation in understanding the nature, purpose, historical context, scientific principles, and ethical considerations that shape clinical research. This knowledge will serve as a solid basis for their further exploration of clinical studies and their roles as participants in the subsequent chapters.

2

THE IMPORTANCE OF CLINICAL RESEARCH

CHAPTER II

THE IMPORTANCE OF CLINICAL RESEARCH

In Chapter 2, we delve deeper into the importance of clinical research in the healthcare field. We will discuss how clinical research plays a vital role in developing new treatments, medications, and medical devices, as well as shaping evidence-based medical practices. You will learn about the impact of clinical research on patient outcomes, healthcare policies, and the overall advancement of medical science.

SECTION 1: ADVANCING MEDICAL KNOWLEDGE

Advancing Medical Knowledge

In this section, we will explore the pivotal role of clinical research in advancing medical knowledge. We will discuss how clinical research provides a systematic and evidence-based approach to understanding diseases, treatments, and healthcare practices. By conducting rigorous studies, researchers can gain insights into the underlying mechanisms of diseases, identify risk factors, and uncover potential interventions that can improve patient outcomes. We will highlight notable discoveries and breakthroughs resulting from clinical research, such as the development of life-saving drugs, innovative surgical techniques, and improved diagnostic tools.

Introduction

Clinical research plays a pivotal role in advancing medical knowledge by providing a systematic and evidence-based approach to understanding diseases, treatments, and healthcare practices. Through rigorous study designs and data analysis, researchers can gain valuable insights that contribute to medical advancements and improvements in patient care. Let's explore how clinical research has led to notable discoveries and breakthroughs in various areas of medicine.

Understanding Disease Mechanisms and Risk Factors

Clinical research enables a deeper understanding of the underlying mechanisms of diseases. Through epidemiological studies, observational research, and clinical trials, researchers can identify risk factors, explore disease progression, and investigate the interactions between genetic, environmental, and lifestyle factors. This knowledge contributes to disease prevention strategies, early detection methods, and personalized medicine approaches.

Development of Life-Saving Drugs

Clinical trials are instrumental in the development and evaluation of new drugs. Through rigorous testing in human subjects, researchers can assess the safety, efficacy, and optimal dosing regimens of pharmaceutical interventions. Notable examples include the discovery of antibiotics like penicillin, which revolutionized the treatment of bacterial infections, and antiretroviral therapy for HIV/AIDS, which transformed the management of the disease and prolonged the lives of millions of individuals.

Innovative Surgical Techniques

Clinical research has paved the way for the development of innovative surgical techniques that have revolutionized various medical specialties. Surgical trials and observational studies have helped refine surgical approaches, improve outcomes, and minimize risks. For instance, the introduction of minimally invasive surgery, such as laparoscopy, has reduced postoperative complications, shortened hospital stays, and improved patient recovery compared to traditional open surgeries.

Improved Diagnostic Tools

Clinical research has led to significant advancements in diagnostic tools and techniques. Through the development and evaluation of new imaging technologies, laboratory tests, and biomarkers, researchers have improved the accuracy and efficiency of disease detection and monitoring. Examples include the use of computed tomography (CT) scans, magnetic resonance imaging (MRI), and molecular diagnostic tests that allow for early detection and precise characterization of diseases.

Prevention and Health Promotion

Clinical research contributes to the identification and validation of preventive measures and health promotion strategies. Studies exploring lifestyle modifications, behavioral interventions, and community-based approaches help inform public health policies and programs. For instance, research on the impact of smoking cessation, exercise, and dietary interventions has played a crucial role in reducing the burden of chronic diseases like cardiovascular disease and cancer.

Advancing Precision Medicine

Precision medicine aims to tailor medical treatments to individual patients based on their unique characteristics, such as genetic makeup, biomarkers, and clinical profiles. Clinical research plays a central role in advancing precision medicine by uncovering the genetic basis of diseases, identifying biomarkers that predict treatment response, and conducting trials to assess the effectiveness of targeted therapies. This approach allows for more personalized and effective treatment strategies, improving patient outcomes and reducing adverse effects.

Conclusion

In summary, clinical research provides a systematic and evidence-based approach to understanding diseases, evaluating treatments, and advancing medical knowledge. Through rigorous studies, researchers gain insights into disease mechanisms, identify risk factors, and uncover potential interventions. This leads to the development of life-saving drugs, innovative surgical techniques, improved diagnostic tools, and advancements in preventive strategies. Clinical research is an indispensable component of medical progress, enabling the delivery of more effective, personalized, and evidence-based care to patients worldwide.

SECTION 2: IMPROVING PATIENT CARE AND TREATMENT OPTIONS

Improving Patient Care and Treatment Options

In this section, we will delve into how clinical research directly benefits patient care. We will emphasize how clinical trials enable the evaluation of new treatments and interventions, ensuring that they are safe and effective before being widely adopted in clinical practice. By participating in clinical research, patients contribute to the development of new therapies and help expand the available treatment options for themselves and future patients. We will provide examples of successful clinical trials that have led to significant advancements in the treatment of various diseases, emphasizing the impact on patient outcomes and quality of life.

Introduction

Clinical research directly benefits patient care by evaluating new treatments and interventions, ensuring their safety and efficacy before they are widely adopted in clinical practice. Clinical trials play a crucial role in this process, and patient participation is vital to the success of these studies. Let's explore how clinical research directly impacts patient care.

Evaluation of New Treatments

Clinical trials are designed to rigorously evaluate the safety and effectiveness of new treatments. These trials follow a predefined protocol that outlines the study objectives, design, procedures, and participant eligibility criteria. By participating in clinical trials, patients have the opportunity to access novel therapies that may offer potential benefits not available through standard treatments. Through their involvement, patients contribute to the generation of scientific evidence that guides treatment decisions and improves patient outcomes.

Ensuring Safety and Efficacy

Clinical trials provide a structured framework to assess the safety and efficacy of new treatments. Prior to conducting trials in humans, extensive preclinical testing is conducted in laboratory settings and animal models. Once a potential treatment shows promise, it progresses to human trials, where its effects are carefully evaluated in a controlled and monitored environment. By closely monitoring participants and collecting data on treatment outcomes, researchers can assess the treatment's safety profile, identify potential side effects, and measure its effectiveness compared to existing options.

Expanding Treatment Options

Clinical research expands the range of available treatment options for patients. By participating in clinical trials, patients gain access to cutting-edge therapies that may not be widely available or accessible through routine clinical practice. This is particularly important for patients with rare diseases, aggressive forms of cancer, or conditions that have limited treatment options. Through clinical research, new therapies are developed and tested, potentially offering improved outcomes or alternative approaches to managing specific health conditions.

Personalized Medicine

Clinical research is advancing the field of personalized medicine, which aims to tailor treatments to individual patients based on their unique characteristics. By participating in clinical trials, patients contribute to the development of targeted therapies that consider factors such as genetic makeup, biomarkers, and clinical profiles. These studies help identify patient subgroups that may respond better to specific treatments, allowing for more precise and individualized care.

Advancements in Standard of Care

Clinical research continually pushes the boundaries of medical knowledge, leading to advancements in the standard of care. Through evidence generated by clinical trials, treatment guidelines and protocols are updated, informing healthcare providers on the most effective and evidence-based approaches to patient care. Research findings also facilitate shared decision-making between patients and healthcare professionals, empowering patients to make informed choices about their treatment options.

Access to Investigational Therapies

In certain cases, patients with life-threatening or debilitating conditions may have limited treatment options available. Clinical trials provide access to investigational therapies that may offer hope and potential benefits. These trials offer an avenue for patients to receive novel treatments before they are approved by regulatory agencies and become widely accessible. In some instances, participation in clinical trials can provide access to therapies that are not yet available through other means.

Continuous Improvement in Patient Care

Clinical research fosters a culture of continuous improvement in patient care. Through the systematic evaluation of treatments and interventions, researchers identify areas for refinement, explore new therapeutic approaches, and strive to enhance patient outcomes. Findings from clinical trials inform medical practice, guide future research endeavors, and ultimately lead to improved standards of care.

Here are a few examples of successful clinical trials that have led to significant advancements in the treatment of various diseases, resulting in improved patient outcomes and quality of life:

1. **HER2-Positive Breast Cancer Treatment (Trastuzumab):** Clinical trials evaluating the effectiveness of trastuzumab, a targeted therapy for HER2-positive breast cancer, have shown remarkable results. These trials demonstrated that adding trastuzumab to standard chemotherapy significantly improves survival rates, reduces the risk of cancer recurrence, and increases the response to treatment. Trastuzumab has become a standard of care for HER2-positive breast cancer, leading to improved outcomes and prolonged survival for many patients.
2. **Antiretroviral Therapy for HIV/AIDS:** Clinical trials evaluating antiretroviral therapy (ART) have revolutionized the management of HIV/AIDS. These trials demonstrated the effectiveness of ART in suppressing the replication of the human immunodeficiency virus (HIV), slowing disease progression, and reducing the risk of opportunistic infections. With the introduction of ART, HIV/AIDS has transformed from a life-threatening disease to a manageable chronic condition, significantly improving the quality and length of life for individuals living with HIV.

3. **Immunotherapy for Advanced Melanoma (Pembrolizumab):** Clinical trials investigating immunotherapy with pembrolizumab have had a transformative impact on the treatment of advanced melanoma, a deadly form of skin cancer. These trials demonstrated that pembrolizumab, a checkpoint inhibitor that boosts the immune system's response against cancer cells, significantly improves overall survival, increases progression-free survival, and produces durable responses in patients with advanced melanoma. Immunotherapy has become a standard treatment option for advanced melanoma, offering new hope and improved outcomes for patients.
4. **Stroke Treatment with Intravenous Tissue Plasminogen Activator (tPA):** Clinical trials evaluating the use of intravenous tissue plasminogen activator (tPA) for acute ischemic stroke have shown its efficacy in restoring blood flow to the brain and reducing disability. These trials demonstrated that early administration of tPA within the therapeutic window improves outcomes, increases the likelihood of a complete recovery, and reduces long-term disability in stroke patients. The use of tPA has become the standard of care for eligible patients with acute ischemic stroke.
5. **Targeted Therapy for Chronic Myeloid Leukemia (Imatinib):**Clinical trials investigating imatinib, a targeted therapy for chronic myeloid leukemia (CML), have had a transformative impact on the treatment of this blood cancer. These trials demonstrated that imatinib, which specifically targets the abnormal protein responsible for CML, induces high rates of complete hematological and cytogenetic remission, prolongs survival, and offers a favorable side effect profile compared to traditional chemotherapy. Imatinib has revolutionized the management of CML, providing long-

term disease control and improved quality of life for patients.

These examples highlight how successful clinical trials have significantly advanced the treatment landscape for various diseases, leading to improved patient outcomes, prolonged survival, and enhanced quality of life. Through rigorous scientific investigation and the participation of patients, these trials have transformed medical practice, offering more effective and targeted therapies that have revolutionized disease management and patient care.

Conclusion

In summary, clinical research directly benefits patient care by evaluating new treatments, ensuring their safety and efficacy, and expanding the range of available treatment options. By participating in clinical trials, patients contribute to the development of new therapies and help advance medical knowledge. Clinical research plays a crucial role in improving patient outcomes, advancing personalized medicine, and continuously enhancing the standard of care for patients worldwide.

SECTION 3: PERSONALIZED MEDICINE AND PRECISION THERAPIES

Personalized Medicine and Precision Therapies

This section will explore the emerging field of personalized medicine and precision therapies, which are revolutionizing patient care. We will discuss how clinical research is at the forefront of developing tailored treatment approaches based on an individual's unique genetic, environmental, and lifestyle factors. We will explain how studies involving genetic sequencing, biomarker identification, and targeted therapies have led to more precise diagnoses, improved treatment responses, and reduced adverse effects. Readers will gain an understanding of how clinical research is driving the shift towards personalized medicine and the potential it holds for optimizing patient outcomes.

Introduction

Personalized medicine, also known as precision medicine, is an emerging field that aims to tailor medical treatment and interventions to individual patients based on their unique genetic, environmental, and lifestyle factors. This approach recognizes that individuals can respond differently to treatments due to variations in their genetic makeup, disease characteristics, and personal circumstances. Clinical research plays a crucial role in advancing personalized medicine by uncovering the genetic basis of diseases, identifying biomarkers, and developing targeted therapies. Let's explore the field of personalized medicine and its impact on patient care in greater detail.

Genetic Understanding of Diseases

Advancements in genetic research have provided insights into the underlying mechanisms of various diseases. Through genome sequencing and genetic profiling, researchers can identify specific genetic alterations associated with certain diseases or treatment responses. Clinical research enables the identification of disease-associated genes, mutations, or variations that help guide treatment decisions and predict disease outcomes. This genetic understanding forms the basis for developing personalized treatment approaches.

Biomarkers and Targeted Therapies

Biomarkers are measurable indicators that provide information about a patient's health status, disease progression, or treatment response. Clinical research aims to identify and validate biomarkers that can guide treatment decisions. By identifying specific biomarkers, such as genetic mutations, protein expression patterns, or molecular signatures, researchers can develop targeted therapies that act on these specific molecular targets. These therapies have the potential to be more effective, less toxic, and have fewer side effects than traditional treatments.

Pharmacogenomics

Pharmacogenomics is a field of study that explores how an individual's genetic makeup influences their response to medications. Through clinical research, researchers identify genetic variations that affect drug metabolism, efficacy, and safety. This information helps healthcare providers tailor drug selection, dosage, and treatment regimens to individual patients, maximizing therapeutic benefits and minimizing adverse effects. Pharmacogenomic testing is increasingly being incorporated into clinical practice to guide medication decisions in areas such as oncology, cardiology, psychiatry, and infectious diseases.

Predictive and Prognostic Models

Clinical research plays a critical role in developing predictive and prognostic models that assist in individualizing treatment decisions. Predictive models use patient-specific information, such as genetic data, biomarkers, and clinical characteristics, to estimate the likelihood of treatment response or adverse events. Prognostic models, on the other hand, help predict disease outcomes and guide treatment planning. These models, often developed through rigorous clinical trials, enable healthcare providers to make informed decisions about the most suitable treatment options for individual patients.

Digital Health and Wearable Devices

Advancements in digital health technologies and wearable devices are revolutionizing personalized medicine. These technologies enable continuous monitoring of patients' health parameters, including heart rate, blood pressure, glucose levels, and activity patterns. Clinical research utilizes these tools to collect real-time data, track disease progression, and assess treatment effectiveness. This information helps personalize treatment plans and allows for timely interventions based on individual patient needs.

Integrating Lifestyle and Environmental Factors

Personalized medicine also considers lifestyle and environmental factors in treatment decisions. Clinical research explores the impact of lifestyle choices, such as diet, exercise, smoking, and stress, on disease development and progression. By integrating this information with genetic and clinical data, researchers can develop personalized strategies that encompass lifestyle modifications alongside targeted therapies. This holistic approach recognizes the importance of addressing individual patient needs and empowering them to actively participate in their own healthcare.

Patient-Centered Care

Personalized medicine puts patients at the center of their healthcare journey. By considering an individual's unique characteristics and preferences, personalized medicine aims to optimize treatment outcomes and improve patient satisfaction. Clinical research ensures that patient perspectives are incorporated into study design, outcome measures, and decision-making processes. Patient-reported outcomes and patient-centered endpoints are increasingly being used to capture the impact of treatments on quality of life and patient well-being.

Continued Advancements in Personalized Medicine

Clinical research involving genetic sequencing, biomarker identification, and targeted therapies has played a transformative role in advancing personalized medicine, leading to more precise diagnoses, improved treatment responses, and reduced adverse effects. Here's a further exploration of these areas:

Genetic Sequencing and Diagnostics

Clinical research has facilitated the development of high-throughput sequencing technologies, such as next-generation sequencing (NGS), which enable the rapid and cost-effective analysis of an individual's genetic material. These advancements have led to the identification of disease-causing genetic mutations, variations, and alterations. Genetic sequencing allows for precise diagnosis and classification of diseases, enabling healthcare providers to stratify patients based on their genetic profiles and select the most appropriate treatment options.

For example, in the field of oncology, genomic profiling of tumors has become increasingly important in guiding treatment decisions. By identifying specific mutations or genetic alterations in cancer cells, clinicians can select targeted therapies that specifically address the underlying genetic drivers of the disease. This approach has led to improved treatment responses and prolonged survival for patients with various types of cancer.

Biomarker Identification and Treatment Response Prediction

Clinical research focuses on identifying and validating biomarkers that can serve as indicators of disease presence, progression, and treatment response. Biomarkers can be genetic, molecular, or cellular characteristics that provide valuable insights into an individual's disease state and treatment needs.

By studying large cohorts of patients and analyzing their genetic and molecular profiles, researchers can identify biomarkers that predict treatment responses. This knowledge allows for the selection of patients who are more likely to benefit from specific interventions, sparing others from potentially ineffective treatments and unnecessary side effects. Biomarkers also enable the development of companion diagnostic tests, which help identify patients who are most likely to respond to a particular therapy.

Targeted Therapies and Precision Treatment Approaches

The identification of specific genetic alterations and biomarkers has paved the way for the development of targeted therapies. These therapies are designed to act on specific molecular targets associated with a particular disease, ensuring more precise and effective treatment.

Targeted therapies can be drugs that directly inhibit the activity of specific proteins or enzymes involved in disease progression. By targeting these specific molecular pathways, targeted therapies can disrupt the growth and survival of cancer cells or modulate the underlying mechanisms of other diseases. Clinical research has demonstrated the efficacy of targeted therapies in improving treatment responses and patient outcomes across various diseases, including cancer, autoimmune disorders.

Conclusion

In summary, personalized medicine and precision therapies are transforming patient care by tailoring treatments to individual patients' genetic, environmental, and lifestyle factors. Clinical research plays a pivotal role in advancing personalized medicine by unraveling the genetic basis of diseases, identifying biomarkers, and developing targeted therapies. Through the integration of genetics, biomarkers, digital health technologies, and patient-centered approaches, personalized medicine holds the promise of improving treatment outcomes, reducing adverse effects, and empowering patients to take an active role in their healthcare.

SECTION 4: ADDRESSING UNMET MEDICAL NEEDS

Addressing Unmet Medical Needs

In this section, we will highlight how clinical research plays a crucial role in addressing unmet medical needs. Many diseases and conditions currently lack effective treatments, leaving patients with limited options. Clinical research provides a platform for testing novel interventions and exploring innovative approaches to tackle these unmet needs. We will discuss the importance of clinical trials in investigating rare diseases, orphan drugs, and conditions that disproportionately affect underserved populations. By participating in clinical research, patients contribute to the development of therapies that may benefit not only themselves but also future generations facing similar medical challenges.

Introduction

Clinical research indeed plays a crucial role in addressing unmet medical needs by providing a platform for testing novel interventions and exploring innovative approaches to tackle diseases and conditions that currently lack effective treatments. Let's dive into the importance of clinical trials in investigating rare diseases, orphan drugs, and conditions that disproportionately affect underserved populations.

Rare Diseases

Rare diseases, also known as orphan diseases, are conditions that affect a small number of individuals. Due to their rarity, there is often limited knowledge about the underlying mechanisms, natural history, and treatment options for these diseases. Clinical research, particularly clinical trials, plays a vital role in advancing the understanding and management of rare diseases.

Clinical trials focusing on rare diseases allow researchers to investigate the safety and efficacy of potential therapies in a controlled setting. They provide an opportunity to test the effectiveness of novel drugs, gene therapies, or other interventions specifically designed to target the underlying causes of the disease. By enrolling patients with rare diseases in clinical trials, researchers can gather valuable data on treatment responses, long-term outcomes, and potential side effects. These trials contribute to the development of evidence-based therapies and improve the quality of care for patients with rare diseases.

Orphan Drugs

Orphan drugs are medications developed specifically for the treatment of rare diseases. These diseases often have limited treatment options, leaving patients with unmet medical needs. Clinical research plays a crucial role in the development and evaluation of orphan drugs.

Clinical trials are essential for assessing the safety and effectiveness of potential orphan drugs. They provide the necessary evidence to support regulatory approvals and subsequent availability of these therapies for patients. Through well-designed clinical trials, researchers can generate data on the drug's efficacy, dosing, potential adverse effects, and long-term outcomes. These trials also contribute to understanding the optimal use of orphan drugs, such as identifying patient populations that are most likely to benefit and determining appropriate dosing regimens.

Underserved Populations

Clinical research has a responsibility to address health disparities and ensure equitable access to effective treatments for all populations, including those that are underserved. Certain diseases, such as certain types of cancers, infectious diseases, and chronic conditions, disproportionately affect underserved communities. Clinical trials offer an opportunity to bridge the gap in healthcare by including diverse populations and addressing the unique needs and challenges they face.

By actively recruiting and involving individuals from underserved populations in clinical research, researchers can ensure that the safety and effectiveness of interventions are assessed in populations that may have been historically underrepresented. This approach helps generate evidence on treatment outcomes and responses within specific populations, facilitating the development of targeted therapies and interventions that consider the diverse needs and characteristics of different communities.

Patient Advocacy and Engagement

Clinical research relies on the participation of patients and their willingness to contribute to the advancement of medical knowledge. By actively engaging with patients and incorporating their perspectives, clinical research becomes more patient-centered and responsive to the unmet needs of specific populations.

Patient advocacy groups and organizations play a vital role in promoting clinical research and supporting patient participation. These groups raise awareness about research opportunities, provide resources for patients considering participation, and advocate for patient rights and protection. Through collaboration between researchers, healthcare professionals, and patient advocacy groups, clinical trials can be designed to address the specific challenges faced by underserved populations and rare disease communities.

Future Generations and Medical Progress

By participating in clinical research, patients contribute not only to their own well-being but also to the advancement of medical knowledge and the development of new therapies. The data and insights generated through clinical trials have a lasting impact on future generations.

Clinical research serves as a foundation for medical progress. The knowledge gained from studying rare diseases, orphan drugs, and underserved populations helps shape future research and treatment strategies. It opens avenues for further exploration, allowing researchers to refine existing therapies, develop new interventions, and improve healthcare outcomes for individuals facing similar medical challenges.

Conclusion

In summary, clinical research plays a critical role in addressing unmet medical needs by investigating rare diseases, orphan drugs, and conditions that disproportionately affect underserved populations. Through clinical trials, researchers can gather crucial data to develop evidence-based therapies, enhance our understanding of disease mechanisms, and improve patient outcomes. By participating in clinical research, patients contribute to the advancement of medical knowledge, the development of new treatments, and the hope for a brighter future in healthcare.

SECTION 5: GUIDING EVIDENCE-BASED MEDICINE

Guiding Evidence-Based Medicine

This section will focus on how clinical research informs evidence-based medicine. We will explain the concept of evidence-based medicine, which emphasizes the integration of clinical expertise, patient values, and the best available research evidence in clinical decision-making. We will highlight the role of well-designed clinical trials and systematic reviews in generating high-quality evidence that guides medical practices. By critically evaluating research findings and applying them in clinical settings, healthcare professionals can provide the most effective and efficient care to their patients.

Introduction

Evidence-based medicine is an approach to healthcare that emphasizes the integration of clinical expertise, patient values, and the best available research evidence in making informed decisions about patient care. It involves critically evaluating and applying scientific evidence to guide medical practices, ensuring that interventions and treatments are based on solid empirical support. Clinical research, particularly well-designed clinical trials and systematic reviews, plays a pivotal role in generating high-quality evidence that informs evidence-based medicine.

Clinical Trials

Well-designed clinical trials are essential in generating robust evidence to guide medical practice and standards of practice. These trials follow rigorous protocols and methodologies to systematically investigate the safety, efficacy, and effectiveness of interventions such as drugs, treatments, and medical devices. Through clinical trials, researchers can compare different interventions, assess their impact on patient outcomes, and identify the most effective approaches.

Randomized controlled trials (RCTs) are considered the gold standard in clinical research. In an RCT, participants are randomly assigned to different intervention groups (e.g., experimental treatment, placebo, or standard care), ensuring that the comparison groups are similar in all aspects except for the intervention. This randomization minimizes bias and allows researchers to draw accurate conclusions about the efficacy of the intervention being studied.

Clinical trials collect data on various outcomes, such as clinical endpoints, quality of life measures, and adverse events. By analyzing these data, researchers can quantify the benefits and risks associated with different interventions, enabling healthcare professionals to make evidence-based decisions when selecting treatments for their patients.

Systematic Reviews and Meta-analyses

Systematic reviews and meta-analyses are critical components of evidence-based medicine. These research methodologies involve the comprehensive and unbiased synthesis of existing research studies on a particular topic or intervention. They aim to provide a summary of the available evidence, identify consistencies or discrepancies among studies, and derive more robust conclusions by pooling data from multiple studies.

Systematic reviews involve a systematic search and appraisal of relevant studies, while meta-analyses employ statistical methods to combine the data from individual studies. By synthesizing the findings of multiple studies, systematic reviews and meta-analyses increase the statistical power and precision of the results, providing a more comprehensive understanding of the effectiveness of interventions.

These synthesis methods allow healthcare professionals to access the collective knowledge from numerous studies, identify patterns or inconsistencies, and make informed decisions based on the overall evidence. They help bridge the gap between research findings and clinical practice by providing clinicians with a consolidated and reliable evidence base.

Critical Appraisal of Research Findings

In evidence-based medicine, healthcare professionals critically appraise research findings to assess the quality, validity, and relevance of the evidence. This involves evaluating the study design, sample size, methodology, statistical analysis, and potential biases in the research.

Critical appraisal helps healthcare professionals determine the strength of the evidence and the applicability of the research findings to their specific patient populations. By critically evaluating research studies, clinicians can differentiate between high-quality evidence and studies with limitations, enabling them to make more informed decisions about patient care.

Integration of Clinical Expertise and Patient Values

Evidence-based medicine recognizes the importance of integrating clinical expertise and patient values into decision-making alongside research evidence. Clinical expertise refers to the knowledge, skills, and experience of healthcare professionals, allowing them to apply research evidence in the context of individual patients' circumstances.

Patient values and preferences are central to the practice of evidence-based medicine. Healthcare professionals consider patient values, beliefs, goals, and preferences when making treatment decisions. They engage in shared decision-making with patients, discussing the available evidence, potential risks and benefits, and considering the patients' preferences in order to arrive at a mutually agreed-upon treatment plan.

By integrating clinical expertise, patient values, and the best available research evidence, healthcare professionals can provide the most effective and efficient care to their patients. Evidence-based medicine aims to optimize patient outcomes and improve the quality of healthcare by ensuring that interventions are based on the best available evidence, tailored to individual patients' needs, and informed by the collective expertise of healthcare professionals.

Conclusion

In conclusion, evidence-based medicine relies on the integration of clinical expertise, patient values, and the best available research evidence. Clinical research, including well-designed clinical trials and systematic reviews, plays a crucial role in generating high-quality evidence. By critically evaluating research findings and applying them in clinical settings, healthcare professionals can make informed decisions and provide optimal care that is grounded in scientific evidence and aligned with patient preferences and values.

SECTION 6: INFLUENCE ON HEALTHCARE POLICIES AND GUIDELINES

Influence on Healthcare Policies and Guidelines

In this section, we will explore how clinical research influences healthcare policies and guidelines. Governments, regulatory bodies, and healthcare organizations rely on robust research evidence to develop policies that shape healthcare practices. We will discuss examples of how clinical research findings have informed the development of screening guidelines, treatment protocols, and public health initiatives. By participating in clinical research, patients contribute to the evidence base that informs these policies, ensuring that healthcare practices align with the latest scientific advancements.

Introduction

Clinical research plays a significant role in influencing healthcare policies and guidelines at various levels, including government agencies, regulatory bodies, and healthcare organizations. These entities rely on robust research evidence to make informed decisions and develop policies that shape healthcare practices. Clinical research findings guide the development of screening guidelines, treatment protocols, and public health initiatives, ensuring that healthcare practices align with the latest scientific advancements. By participating in clinical research, patients contribute to the evidence base that informs these policies, ultimately improving healthcare delivery and patient outcomes.

Development of Screening Guidelines

Screening guidelines aim to identify individuals at risk for specific diseases or conditions so that early intervention can be initiated. Clinical research provides essential evidence for the development of screening guidelines by determining the effectiveness, benefits, and potential harms of various screening methods.

For example, research studies have informed the development of breast cancer screening guidelines recommending regular mammograms for women of certain age groups. The evidence from clinical trials and observational studies has helped establish the optimal age to initiate screening, the frequency of screening, and the effectiveness of different screening modalities. These guidelines enable healthcare providers to identify and detect breast cancer at early stages, leading to improved outcomes and reduced mortality rates.

Treatment Protocols and Clinical Practice Guidelines

Clinical research findings influence the development of treatment protocols and clinical practice guidelines, which provide healthcare professionals with evidence-based recommendations for managing specific diseases or conditions. These guidelines ensure consistent, high-quality care across healthcare settings and assist healthcare professionals in making informed treatment decisions.

For instance, clinical trials comparing different treatment approaches for hypertension (high blood pressure) have influenced the development of treatment protocols and guidelines. Research findings have provided evidence on the effectiveness of specific medications, optimal blood pressure targets, and the benefits of lifestyle modifications. These guidelines serve as a reference for healthcare providers in selecting appropriate treatments and monitoring patient progress, leading to improved management of hypertension and reduced associated complications.

Public Health Initiatives and Policy Decisions

Clinical research also informs public health initiatives and policy decisions aimed at promoting population health and preventing diseases. By identifying risk factors, evaluating interventions, and measuring outcomes, clinical research contributes to evidence-based public health strategies.

For example, research studies on the effectiveness of vaccinations have influenced public health policies regarding immunization schedules, vaccine safety, and vaccine coverage rates. Clinical trials and epidemiological studies have provided evidence on vaccine efficacy, duration of protection, and the impact of vaccination on disease transmission. These findings have informed policies that guide immunization programs, leading to significant reductions in vaccine-preventable diseases and improved population health.

Regulation and Drug Approval

Clinical research is fundamental to the regulation and approval of new drugs and medical interventions. Regulatory bodies, such as the Food and Drug Administration (FDA) in the United States and the European Medicines Agency (EMA) in Europe, rely on rigorous clinical trials to evaluate the safety and efficacy of new treatments before they can be made available to the public.

Clinical trials provide evidence of a treatment's benefits, potential risks, and side effects, helping regulatory bodies make informed decisions regarding drug approvals and labeling. These regulatory decisions ensure that new treatments meet rigorous standards of safety and effectiveness, protecting patient welfare and promoting public health.

Conclusion

In summary, clinical research findings have a profound impact on healthcare policies and guidelines. Governments, regulatory bodies, and healthcare organizations rely on robust research evidence to develop screening guidelines, treatment protocols, and public health initiatives. By participating in clinical research, patients contribute to the evidence base that informs these policies, ensuring that healthcare practices align with the latest scientific advancements. Ultimately, clinical research plays a crucial role in shaping healthcare policies, improving healthcare delivery, and enhancing patient outcomes.

SECTION 7: COLLABORATIONS AND MULTIDISCIPLINARY APPROACHES

Collaborations and Multidisciplinary Approaches

This section will highlight the collaborative nature of clinical research. We will emphasize how clinical research brings together researchers, healthcare professionals, industry stakeholders, patients, and advocacy groups to collectively advance medical knowledge. We will discuss the importance of multidisciplinary approaches, where experts from various fields collaborate to address complex research questions and develop innovative solutions. Readers will gain an understanding of the diverse expertise and teamwork required to conduct successful clinical research studies.

Introduction

Clinical research is a highly collaborative endeavor that brings together researchers, healthcare professionals, industry stakeholders, patients, and advocacy groups with the shared goal of advancing medical knowledge and improving patient care. The collaborative nature of clinical research is rooted in the recognition that addressing complex research questions and developing innovative solutions requires diverse expertise and multidisciplinary approaches.

Researchers

Clinical researchers, including scientists, physicians, and other healthcare professionals, play a central role in designing, conducting, and analyzing clinical research studies. They bring their scientific knowledge, research skills, and clinical experience to the table, ensuring that studies are methodologically sound, ethical, and relevant to clinical practice. Researchers collaborate with colleagues within their own institutions, as well as with external collaborators and research networks, to leverage expertise and resources.

Healthcare Professionals

Healthcare professionals, such as physicians, nurses, and allied health providers, contribute invaluable clinical expertise and patient perspectives to the research process. They often serve as principal investigators or co-investigators in clinical studies, contributing to study design, participant recruitment, data collection, and interpretation of results. Healthcare professionals involved in clinical research are uniquely positioned to bridge the gap between research and clinical practice, translating research findings into evidence-based care for their patients.

Industry Stakeholders

Pharmaceutical and medical device companies, biotechnology firms, and other industry stakeholders play a critical role in clinical research. They often sponsor and fund clinical trials, providing the necessary resources, infrastructure, and expertise to bring investigational products to market. Collaborating with industry partners allows researchers to access novel therapies, study interventions, and cutting-edge technologies. These partnerships facilitate the translation of scientific discoveries into tangible healthcare advancements.

Patient and Advocacy Groups

Patients are vital partners in clinical research. Their participation and input are essential for the development of patient-centered research questions, study design, and outcome measures. Patient advocacy groups represent the interests and perspectives of specific patient populations, helping to shape research priorities, raise awareness, and facilitate patient engagement in research. Engaging patients and advocacy groups ensures that research is relevant, respectful, and addresses the needs and concerns of the individuals it aims to benefit.

Multidisciplinary Approaches

Clinical research often requires collaboration across multiple disciplines to address complex questions and challenges. Experts from various fields, such as genetics, epidemiology, biostatistics, psychology, pharmacology, and health economics, contribute their specialized knowledge and skills to the research process. For example, a study investigating the effectiveness of a new cancer treatment may involve collaboration between oncologists, pathologists, molecular biologists, statisticians, and patient advocates. This multidisciplinary approach fosters comprehensive and holistic research, enhancing the validity and impact of the findings.

Successful clinical research studies depend on effective communication, teamwork, and shared decision-making among the diverse stakeholders involved. Collaboration occurs at various stages of the research process, from study design and protocol development to data analysis, interpretation, and dissemination of results. Regular meetings, research conferences, and collaborative platforms facilitate information exchange, foster dialogue, and promote collective learning.

Moreover, collaborative clinical research efforts extend beyond individual studies. Research networks, consortia, and international collaborations enable the pooling of data and resources, allowing for larger sample sizes, more diverse populations, and enhanced generalizability of findings. These collaborative initiatives promote transparency, knowledge sharing, and the acceleration of scientific discoveries.

Conclusion

In conclusion, clinical research is a highly collaborative undertaking that brings together researchers, healthcare professionals, industry stakeholders, patients, and advocacy groups. This collective effort leverages diverse expertise, multidisciplinary approaches, and shared resources to advance medical knowledge, improve patient care, and address complex research questions. By working together, these stakeholders contribute to the success of clinical research studies and ultimately benefit patients and society as a whole.

SECTION 8: GLOBAL HEALTH AND CLINICAL RESEARCH

Global Health and Clinical Research

In this section, we will explore the global impact of clinical research on public health. We will discuss how clinical research contributes to global health initiatives, such as the development of vaccines, the study of infectious diseases, and the evaluation of interventions to address global health challenges. We will highlight the importance of conducting research in diverse populations to ensure the generalizability of findings and address health disparities on a global scale.

Introduction

Clinical research has a significant global impact on public health by contributing to the development of vaccines, studying infectious diseases, and evaluating interventions to address global health challenges. Clinical research plays a crucial role in advancing our understanding of diseases and healthcare practices, with the ultimate goal of improving health outcomes for populations around the world.

Development of Vaccines

Vaccines have had a profound impact on global health by preventing the spread of infectious diseases and reducing associated morbidity and mortality. Clinical research plays a pivotal role in vaccine development, including preclinical studies, safety and efficacy trials, and post-marketing surveillance. Clinical trials are conducted to assess the safety, immunogenicity, and effectiveness of vaccines in diverse populations. These trials provide the evidence needed for regulatory approval and inform global immunization policies.

For example, clinical research played a crucial role in the development of vaccines against major infectious diseases such as polio, measles, tuberculosis, and hepatitis. Through rigorous clinical trials, these vaccines have been proven safe and effective, leading to their widespread implementation and significant reductions in disease burden worldwide.

Study of Infectious Diseases

Clinical research is essential for understanding the transmission, pathogenesis, treatment, and prevention of infectious diseases that affect global populations. By conducting clinical trials and observational studies, researchers gain insights into the natural history of diseases, identify risk factors, evaluate diagnostic tools, and test novel therapies.

For instance, clinical research has been instrumental in advancing the understanding and management of HIV/AIDS. Through large-scale clinical trials, antiretroviral therapies have been developed, evaluated, and optimized, leading to significant improvements in the quality of life and life expectancy of individuals living with HIV/AIDS. Clinical research also informs preventive strategies such as pre-exposure prophylaxis (PrEP), which has proven effective in reducing the risk of HIV transmission.

Evaluation of Interventions for Global Health Challenges

Clinical research plays a vital role in evaluating interventions to address global health challenges, including non-communicable diseases, maternal and child health, and healthcare delivery in resource-limited settings. By conducting clinical trials and implementation studies, researchers gather evidence on the effectiveness, feasibility, and scalability of interventions.

For example, clinical research has been instrumental in advancing maternal and child health outcomes globally. Studies have evaluated interventions to reduce maternal mortality, improve prenatal care, enhance neonatal health, and promote childhood immunization. These studies provide evidence-based recommendations that guide policy and programmatic interventions, leading to improved health outcomes for women and children worldwide.

Diverse Populations and Health Disparities

To ensure the generalizability of research findings and address health disparities, clinical research must include diverse populations from various geographic locations, ethnic backgrounds, and socioeconomic strata. By including underrepresented populations in clinical studies, researchers gain insights into the unique health needs and responses to interventions in different populations.

Inclusion of diverse populations in clinical research is crucial for addressing health disparities and ensuring equitable access to healthcare. Research studies have highlighted disparities in disease incidence, treatment response, and health outcomes among different populations. By identifying and understanding these disparities, clinical research can inform interventions and policies aimed at reducing health inequities globally.

In summary, clinical research has a profound impact on global public health. It contributes to the development of vaccines, the study of infectious diseases, and the evaluation of interventions to address global health challenges. By conducting research in diverse populations, clinical research ensures the generalizability of findings and helps address health disparities on a global scale. The knowledge gained from clinical research enables the implementation of evidence-based strategies, leading to improved health outcomes and the achievement of global health initiatives.

Wrapping Up

By the end of Chapter 2, readers will have a comprehensive understanding of the immense importance of clinical research. They will recognize how clinical research advances medical knowledge, improves patient care, guides evidence-based medicine, influences healthcare policies, and contributes to global health initiatives. This understanding will underscore the significance of their participation as research subjects and inspire them to support and advocate for clinical research.

3

TYPES OF CLINICAL RESEARCH STUDIES

CHAPTER III

TYPES OF CLINICAL RESEARCH STUDIES

Clinical research is a multifaceted field, distinguished by the diversity of study designs that researchers can employ to explore, understand, and ultimately improve health outcomes. In this chapter, we will delve into the various types of clinical research studies, each tailored to answer specific research questions underpinning health interventions. This chapter serves as a comprehensive guide to understanding the distinctions, applications, and implications of each study type within the broader context of clinical research.

SECTION 1: INTERVENTIONAL STUDIES

Interventional Studies

This section will focus on interventional studies, which involve the deliberate manipulation of variables to evaluate the effectiveness of specific interventions. We will discuss the purpose of interventional studies, which is to assess the safety, efficacy, and optimal use of treatments, medications, procedures, or behavioral interventions. We will explain the different types of interventional studies, including randomized controlled trials (RCTs), non-randomized controlled trials, and single-arm studies. Readers will gain an understanding of how interventional studies provide robust evidence for the effectiveness of interventions and contribute to clinical practice guidelines.

Introduction

Interventional studies are a crucial component of clinical research that involve the deliberate manipulation of variables to evaluate the safety, efficacy, and optimal use of specific interventions. These studies aim to provide robust evidence regarding the effectiveness of treatments, medications, procedures, or behavioral interventions in real-world settings. By actively intervening and comparing different approaches, interventional studies help guide clinical decision-making and contribute to the development of evidence-based guidelines.

Purpose of Interventional Studies

The primary purpose of interventional studies is to assess the effects of specific interventions on patient outcomes. These studies are designed to determine whether an intervention is safe, effective, and superior to existing treatment options or placebo. Interventional studies also investigate the optimal dosage, duration, and administration of interventions to optimize outcomes.

Types of Interventional Studies

Randomized Controlled Trials (RCTs)

RCTs are considered the gold standard for evaluating the effectiveness of interventions. Participants in an RCT are randomly assigned to different groups: the experimental group receiving the intervention being tested and the control group receiving a placebo or standard treatment. Randomization minimizes selection bias and ensures that potential confounding factors are equally distributed between the groups. RCTs often employ blinding techniques (single-blind or double-blind) to minimize bias. RCTs provide strong evidence for causality and are widely used in clinical research.

Non-Randomized Controlled Trials

In non-randomized controlled trials, participants are assigned to different groups based on non-random processes, such as patient choice or clinical judgment. While these studies lack the strict randomization of RCTs, they can still provide valuable evidence when randomization is not feasible or ethically acceptable. Statistical techniques, such as propensity score matching or adjustment, are used to control for potential confounding factors.

Single-Arm Studies

Single-arm studies evaluate the effectiveness of an intervention without a control group. Participants receive the intervention, and outcomes are assessed based on pre-defined measures. Single-arm studies are typically used in situations where it is not ethical or feasible to include a control group, such as when testing a potentially life-saving intervention. While single-arm studies provide limited evidence on the effectiveness of interventions, they can be valuable for generating hypotheses and exploring initial safety and efficacy data.

Advantages of Interventional Studies

1. **Causality and Treatment Effects:** Interventional studies, particularly RCTs, provide robust evidence for establishing causality and determining the true treatment effects of interventions. By comparing different groups, researchers can directly attribute observed outcomes to the interventions being tested.
2. **Rigorous Design and Control:** Interventional studies employ rigorous design, randomization, blinding, and control measures to minimize bias and confounding

factors. This enhances the internal validity of the study and strengthens the reliability of the findings.

3. **Standardization:** Interventional studies often involve standardized protocols for intervention administration, outcome assessment, and data collection, ensuring consistency and comparability across different study sites.
4. **Regulatory Approval:** Well-conducted interventional studies provide critical data for obtaining regulatory approval for new interventions, such as medications or medical devices.

Limitations of Interventional Studies

1. **Ethical Considerations:** Some interventions may raise ethical concerns when used in a controlled trial, especially if they are already considered standard of care. Ethical considerations may limit the use of placebo or control groups in certain situations.
2. **External Validity:** The controlled nature of interventional studies may limit the generalizability of the findings to real-world settings. Strict eligibility criteria and specific study conditions may not fully represent the broader population.
3. **Time and Resource Intensive:** Interventional studies require significant time, resources, and coordination to design, implement, and analyze. They often involve large sample sizes, long-term follow-ups, and multiple study sites, making them complex and costly endeavors.
4. **Potential Biases:** Despite rigorous design and control measures, interventional studies may still be subject to biases, such as selection bias, attrition bias, or performance bias. Careful study design and statistical analysis help mitigate these biases.

Conclusion

In conclusion, interventional studies play a critical role in clinical research by assessing the safety, efficacy, and optimal use of interventions. Randomized controlled trials are considered the gold standard for evaluating interventions, but non-randomized controlled trials and single-arm studies also contribute valuable evidence. Interventional studies provide robust evidence for clinical practice guidelines, help inform treatment decisions, and contribute to the advancement of medical knowledge. However, they also have limitations, and careful consideration of study design, ethical considerations, and potential biases is necessary to ensure the validity and generalizability of the findings.

SECTION 2: EXPERIMENTAL DESIGNS IN CLINICAL RESEARCH

Experimental Designs in Clinical Research

In this section, we will delve deeper into experimental designs commonly used in clinical research. We will discuss the key features of randomized controlled trials (RCTs), including randomization, blinding, and the use of control groups. We will explain the advantages of RCTs in minimizing bias and confounding variables, thereby providing strong evidence for causal relationships. Additionally, we will discuss other experimental designs, such as factorial designs and crossover designs, and their applications in clinical research. Readers will gain insight into the rigorous methodology employed in experimental designs and understand their importance in generating reliable and valid results.

Experimental designs play a crucial role in clinical research, providing a rigorous framework for investigating the effectiveness of interventions and establishing causal relationships. Among the various experimental designs, randomized controlled trials (RCTs) are widely regarded as the gold standard due to their robust methodology and ability to minimize bias and confounding variables.

Randomized Controlled Trials (RCTs)

RCTs are designed to evaluate the effectiveness of interventions by randomly assigning participants to different groups: the experimental group receiving the intervention being tested and the control group receiving either a placebo or standard treatment. Randomization ensures that participants have an equal chance of being assigned to either group, minimizing selection bias and creating comparable groups at baseline. RCTs typically employ blinding techniques to further minimize bias.

Randomization

Randomization ensures that potential confounding factors are distributed equally between the groups, allowing for valid comparisons. Random assignment helps control for both known and unknown factors that could influence the outcomes, reducing the risk of confounding.

Blinding

Blinding, also known as masking, is employed to minimize bias in RCTs. In single-blind studies, participants are unaware of their assigned group, while in double-blind studies, both participants and researchers are unaware. Blinding helps prevent conscious or unconscious bias in outcome assessment and data interpretation.

Control Group

The control group provides a comparison against which the intervention group is evaluated. Control groups can receive a placebo, standard treatment, or an active comparator. By comparing the outcomes between the groups, researchers can determine the true effect of the intervention being tested.

Advantages of RCTs

1. **Minimization of Bias:** RCTs are designed to minimize bias in the allocation of interventions. By randomly assigning participants to either the treatment group or the control group, the trials ensure that each group is similar in all respects at the start of the experiment. This randomization helps balance both known and unknown confounding variables across the groups, which might otherwise influence the outcome and skew the results. The elimination of selection bias increases the likelihood that differences in outcomes are truly due to the intervention itself, not to differences between participant groups.
2. **Causality:** One of the most significant strengths of RCTs is their ability to demonstrate causality. Because the only expected difference between the treatment and control groups is the intervention itself, any differences in outcomes can be attributed directly to the intervention with a high degree of confidence. This is crucial for assessing the efficacy and safety of new drugs, therapies, or medical devices, and is less reliably achieved with other study designs.
3. **Statistical Power:** RCTs are typically structured to include statistical power calculations that ensure a sufficient sample size to detect a clinically meaningful effect of the intervention, should one exist. This methodical approach to study design and data analysis means that RCTs can provide definitive conclusions about the effectiveness of an intervention. Moreover, well-planned RCTs can include multiple endpoints or outcomes to address various research questions simultaneously, maximizing the utility and impact of the research.

These advantages make RCTs a powerful tool in clinical research, especially when the goal is to assess the direct impact of interventions on health outcomes under controlled and rigorously monitored conditions.

Limitations of RCTs

1. **Generalizability:** While RCTs are excellent for determining causality within the controlled environment of the trial, their strict inclusion and exclusion criteria can limit the generalizability of the findings to the broader population. Participants in RCTs often do not represent the full spectrum of patients who might use the medical product in real-world settings. For example, older adults, children, pregnant women, or individuals with multiple chronic conditions are frequently excluded from trials.
2. **Cost and Time:** RCTs are often expensive and time-consuming to conduct. They require significant resources in terms of funding, staffing, and infrastructure. The need to recruit sufficient participants who meet specific criteria, follow rigorous protocols, and conduct long-term follow-up for outcomes can be resource intensive. This makes it challenging to initiate and sustain RCTs, especially for conditions that are rare or where the expected effect size is small.
3. **Ethical Concerns:** There are ethical considerations in conducting RCTs, particularly when withholding potentially beneficial treatments from the control group might be necessary to assess the treatment's efficacy. This poses ethical dilemmas, especially in trials involving serious or life-threatening conditions where existing treatments are inadequate or have significant side effects. Ensuring ethical standards can sometimes require complex trial designs, such as crossover trials or the use of

placebo only when no proven effective treatment exists, which can further complicate the study design and interpretation of results.

Understanding these limitations is crucial for interpreting the results of RCTs and for planning future research that addresses these challenges while maintaining the rigor and benefits that RCTs provide.

Factorial Designs

Factorial designs involve the simultaneous evaluation of two or more interventions or factors. This design allows for the assessment of main effects (the effects of individual interventions) and interaction effects (the combined effects of interventions). By manipulating multiple variables, researchers can explore the independent and combined effects of interventions. Factorial designs can increase the efficiency of the study by evaluating multiple interventions simultaneously, reducing the sample size requirements compared to conducting separate studies.

Crossover Designs

Crossover designs involve multiple treatment periods, where each participant receives different interventions in a predetermined sequence. Participants act as their control, as they serve as both the experimental and control group at different time points. Crossover designs are commonly used when studying chronic conditions or interventions with reversible effects. This design allows for within-subject comparisons, reducing inter-subject variability and increasing the statistical power of the study.

Conclusion

Experimental designs, particularly RCTs, provide a strong foundation for generating reliable and valid results in clinical research. By incorporating randomization, blinding, and control groups, RCTs minimize bias and confounding variables, enabling researchers to draw robust conclusions about the effectiveness of interventions. Factorial designs and crossover designs offer additional flexibility and efficiency in studying multiple interventions or evaluating within-subject comparisons. These rigorous experimental designs contribute to the advancement of medical knowledge and aid in evidence-based decision-making in clinical practice.

SECTION 3: NON-INTERVENTIONAL STUDIES

Non-Interventional Studies

This section will focus on non-interventional studies, also known as observational or non-experimental studies. We will explain that non-interventional studies primarily aim to gather data and observe outcomes without any intervention or manipulation of variables. We will discuss the various types of non-interventional studies, such as surveys, registries, and case series. Readers will understand how non-interventional studies provide valuable real-world evidence on the effectiveness, safety, and utilization of interventions in routine clinical practice. We will also discuss the limitations of non-interventional studies, including potential biases and confounding factors.

Introduction

Non-interventional studies, also known as observational or non-experimental studies, are a fundamental type of clinical research that aims to observe and collect data on participants in their natural settings without intervening or manipulating variables. Unlike interventional studies, non-interventional studies do not involve any deliberate manipulation of variables or interventions. Instead, they aim to observe and gather data on participants in their natural settings, allowing researchers to investigate real-world outcomes, patterns, and associations. These studies play a crucial role in exploring associations, identifying risk factors, and understanding the natural history of diseases. By observing participants over time, researchers can gain valuable insights into various aspects of health and disease without the constraints of a controlled experimental setting.

Types of Non-Interventional Studies

Surveys

Surveys are a common type of non-interventional study that involve collecting information from participants through questionnaires or interviews. Surveys can provide valuable insights into participants' perspectives, behaviors, and experiences related to their health, treatment choices, or quality of life. They are particularly useful for gathering data on subjective measures, such as patient-reported outcomes and satisfaction.

Registries

Registries are databases that systematically collect data on individuals with a particular disease, condition, or exposure. They can be disease-specific or encompass broader populations. Registries capture information on demographics, medical history, treatments, and outcomes over time. By analyzing registry data, researchers can assess the effectiveness and safety of interventions in real-world settings, monitor disease progression, identify prognostic factors, and generate hypotheses for further investigation.

Cohort Studies

Cohort studies follow a group of individuals with a common characteristic or exposure over a specific period. Researchers collect data on participants' baseline characteristics and follow them prospectively to observe outcomes of interest. Cohort studies can be either prospective (following participants into the future) or retrospective (analyzing historical data). They allow for the examination of the association between exposures or risk factors and outcomes, providing valuable insights into the natural history of diseases and the identification of risk or protective factors.

Case-Control Studies

In case-control studies, researchers identify individuals with a specific outcome (cases) and compare them to individuals without the outcome (controls). The study then assesses the exposure history of both groups to determine if there is an association between exposures and the outcome. Case-control studies are useful for investigating rare outcomes or diseases with long latency periods. They are particularly valuable when studying diseases with long latency periods or when it is difficult to establish a prospective cohort.

Cross-Sectional Studies

Cross-sectional studies collect data on exposure and outcome simultaneously at a specific point in time. These studies provide a snapshot of the prevalence of exposures and outcomes within a population. Cross-sectional studies are often used to assess the burden of diseases, identify risk factors, or explore associations between variables. They are useful for generating hypotheses, estimating the burden of a disease or condition, and identifying associations between variables, but cannot establish causality or examine temporal relationships.

Advantages of Non-Interventional Studies

1. **Real-World Context:** By observing participants in their natural settings, non-interventional studies provide insights into how interventions and treatments are used, their effectiveness, and their impact on real-world populations and patient outcomes in routine clinical practice.
2. **Large Sample Sizes:** Non-interventional studies often involve large sample sizes, allowing for the examination of rare outcomes or subgroups that may be underrepresented in interventional studies. This increases statistical power and allows for the examination of rare outcomes or subgroup analyses.
3. **Longitudinal Perspective:** By following participants over time, non-interventional studies allow researchers to assess the temporal relationship between exposures and outcomes. This temporal perspective is valuable for establishing causality and understanding the natural history of diseases.
4. **Generalizability:** Findings from non-interventional studies are often more generalizable to real-world populations and clinical practice than those from highly controlled interventional studies.
5. **Cost-Effectiveness:** Non-interventional studies are often more cost-effective and require less time than randomized controlled trials, making them more feasible for studying long-term outcomes or rare diseases.

Limitations of Non-Interventional Studies

1. **Confounding and Bias:** Since participants are not randomly assigned to exposures, observational studies are susceptible to confounding factors that may influence the relationship between exposure and outcome. Researchers employ various statistical techniques, such as multivariable regression or propensity score matching, to control for confounding factors. These factors can affect the validity and interpretation of the results.
2. **Lack of Randomization:** Without randomization, non-interventional studies cannot establish causality or determine the true effectiveness of interventions. The presence of unmeasured confounders or biases may affect the observed associations.
3. **Limited Ability to Establish Temporality:** Non-interventional studies cannot establish a temporal relationship between exposures and outcomes as effectively as experimental designs. They rely on existing data or observations rather than manipulating variables or randomization.
4. **Recall Bias and Misclassification:** In non-interventional studies that rely on self-reported data, participants may have recall bias, leading to inaccuracies in exposure or outcome reporting. Misclassification of exposures or outcomes can also occur, affecting the validity of the results.
5. **Limited Control over Variables:** In non-interventional studies, researchers do not have control over the variables being measured. This lack of control may limit the ability to draw definitive conclusions about causality or isolate the specific effects of an exposure.

Conclusion

Non-interventional studies play a crucial role in generating real-world evidence, providing insights into the effectiveness, safety, and utilization of interventions in routine clinical practice. By capturing data from diverse populations and observing outcomes in natural settings, these studies complement the findings from interventional studies, offering a comprehensive understanding of healthcare interventions and their impact on patient outcomes.

SECTION 4: QUALITATIVE RESEARCH

Qualitative Research

In this section, we will introduce qualitative research, a type of research that focuses on understanding human behavior, experiences, and perceptions. We will explain the purpose of qualitative research, which is to explore subjective experiences, gather in-depth insights, and generate rich narratives. We will discuss common qualitative research methods, such as interviews, focus groups, and ethnographic observations. Readers will gain an understanding of how qualitative research complements quantitative research by providing a deeper understanding of the human aspects of healthcare and generating hypotheses for further exploration.

Introduction

Qualitative research is a valuable approach in clinical research that aims to understand and explore the complexities of human behavior, experiences, and perceptions. It focuses on gathering in-depth insights and generating rich narratives to uncover the subjective meaning individuals attach to their experiences. Unlike quantitative research, which relies on numerical data and statistical analysis, qualitative research emphasizes the depth and context of human experiences.

The purpose of qualitative research is to gain a comprehensive understanding of the human aspects of healthcare, such as patients' perspectives, healthcare provider experiences, and the social and cultural influences on healthcare delivery. By exploring subjective experiences, beliefs, and motivations, qualitative research helps uncover insights that quantitative data alone cannot capture. It also generates hypotheses and theoretical frameworks for further investigation.

Common Qualitative Research methods include:

Interviews

Interviews are a key method in qualitative research, allowing researchers to have in-depth conversations with individuals or groups. In individual interviews, researchers ask open-ended questions and encourage participants to share their experiences, beliefs, and perspectives. Group interviews, also known as focus groups, involve facilitating discussions among a small group of participants, enabling the exploration of shared experiences and group dynamics.

Ethnographic Observations

Ethnographic observations involve immersing researchers in the natural environment of the participants to understand their behavior and experiences. Researchers observe and document social interactions, cultural practices, and contextual factors that influence healthcare delivery. Ethnographic research provides rich contextual data, highlighting the social and cultural dimensions that shape healthcare experiences.

Content Analysis

Content analysis involves systematically analyzing written, visual, or audiovisual materials such as documents, transcripts, photographs, or videos. Researchers identify themes, patterns, and meanings within the data, enabling the exploration of commonalities and variations in participants' experiences or representations of healthcare-related phenomena.

Narrative Analysis

Narrative analysis focuses on understanding the stories and narratives shared by individuals. Researchers examine the structure, content, and context of narratives to identify common themes, plotlines, and characterizations. This approach helps uncover the personal and subjective meanings individuals attribute to their experiences, emphasizing the importance of storytelling in understanding human experiences.

Advantages of Qualitative Research

1. **Depth and Detail:** Qualitative research provides a depth of understanding and detail that is not usually possible with quantitative methods. By using open-ended questions, observations, and interviews, researchers can gather rich insights into human behaviors, motivations, emotions, and the underlying reasons behind people's actions. This approach allows for a comprehensive exploration of complex phenomena.
2. **Contextual Understanding:** One of the major strengths of qualitative research is its ability to capture data about the contexts and social settings in which participants operate. Researchers can observe and interact with participants in their natural environments, which helps in understanding the influences of contextual factors on behaviors and experiences. This context-rich data provides a more holistic view of the subject matter.
3. **Flexibility:** Qualitative research methods are often more flexible compared to quantitative methods. Researchers can adjust their approaches based on initial findings to probe deeper into the issue or to explore unexpected insights. This adaptability can lead to more comprehensive and tailored insights that rigid, predefined quantitative methods might miss.
4. **Development of Initial Insights and Hypotheses:** Qualitative research is particularly useful in the exploratory phases of studies where the goal is to generate hypotheses or theories rather than testing them. It can uncover trends, themes, and patterns that inform later quantitative research, providing a solid foundation for further study. This is particularly valuable in new fields of study or areas where little is known.

5. **Human-centered and Subjective Insights:** Qualitative research emphasizes subjective experiences and often seeks to understand phenomena from the perspective of the research participants themselves. This human-centered approach can reveal the meanings people attach to their choices and actions, which is especially important in fields like psychology, sociology, and user experience research.

These advantages make qualitative research a powerful tool in understanding complex, nuanced aspects of human life and social interactions that are not easily quantified or generalized but are essential for both theoretical and applied knowledge.

Limitations of Qualitative Research

1. **Subjectivity:** One of the major criticisms of qualitative research is its potential for subjectivity. The presence and perceptions of the researcher can influence all aspects of the study, from the framing of the question to the collection and interpretation of data. This can lead to bias and affect the reliability and objectivity of the findings, making it difficult to separate the researcher's personal biases from the observed facts.
2. **Generalizability:** Due to typically smaller sample sizes and the non-random selection of participants, the findings from qualitative research are often not generalizable to larger populations. This limited generalizability makes it difficult to apply the results broadly or to predict wider trends with high confidence.
3. **Reproducibility:** Qualitative data collection methods such as interviews and focus groups are dynamic, with the potential for questions and discussions to vary significantly across different sessions. This variability can make it challenging to replicate the study exactly, which is a key criterion for validating research results in the scientific community.
4. **Time and Resource Intensity:** Conducting qualitative research can be resource-intensive and time-consuming. Gathering data through methods like prolonged observation, in-depth interviews, or extensive fieldwork requires substantial time investment both from researchers and participants. Additionally, analyzing qualitative data manually is often a meticulous and lengthy process that demands significant effort and expertise.
5. **Data Overload:** Qualitative research often generates large volumes of data, which can be overwhelming to analyze

comprehensively. The richness of the data, while a strength, can also be a limitation as researchers may find it challenging to systematically organize and interpret extensive textual or visual information without losing valuable insights.

Despite these limitations, qualitative research remains invaluable for exploring complex, nuanced issues in depth, particularly when quantitative methods might fail to capture the full spectrum of human experiences and social dynamics.

Conclusion

In conclusion, qualitative research complements quantitative research by providing a deeper understanding of the human aspects of healthcare. Through methods such as interviews, focus groups, and ethnographic observations, qualitative research uncovers rich narratives, explores subjective experiences, and generates hypotheses for further exploration. Its focus on context, meaning, and social influences enhances our understanding of healthcare experiences and can inform the development of patient-centered interventions and policies.

SECTION 5: SYSTEMATIC REVIEWS AND META-ANALYSES

Systematic Reviews and Meta-analyses

This section will highlight the importance of systematic reviews and meta-analyses in clinical research. We will explain that systematic reviews involve a rigorous process of identifying, appraising, and synthesizing existing research evidence on a specific topic. We will discuss the benefits of systematic reviews in providing a comprehensive and unbiased summary of the available evidence. Additionally, we will explain how meta-analyses combine data from multiple studies to obtain a more precise estimate of the intervention effect. Readers will understand how systematic reviews and meta-analyses inform clinical decision-making by providing a higher level of evidence.

Introduction

Systematic reviews and meta-analyses play a crucial role in clinical research by providing a comprehensive and rigorous synthesis of existing evidence. They are powerful tools for summarizing and analyzing the results of multiple studies on a specific research question or topic. Systematic reviews involve a meticulous and transparent process of identifying, critically appraising, and synthesizing relevant studies, while meta-analysis combines the quantitative data from these studies to obtain a more precise estimate of the intervention effect.

The importance of systematic reviews lies in their ability to provide a high level of evidence for clinical decision-making. Let's cover and discuss some key aspects considering systematic reviews.

Comprehensive and Unbiased Summary

Systematic reviews systematically search and include all relevant studies, minimizing bias and ensuring a comprehensive overview of the available evidence. By adhering to pre-defined criteria and rigorous methodologies, systematic reviews reduce the risk of selective inclusion or exclusion of studies, leading to a more balanced summary.

Appraisal of Study Quality

Systematic reviews critically appraise the quality and methodological rigor of included studies. This evaluation helps identify potential biases, limitations, and sources of heterogeneity across studies. By considering study quality, systematic reviews provide a critical assessment of the strength of the evidence.

Synthesis of Findings

Systematic reviews synthesize the findings of multiple studies, allowing researchers and clinicians to grasp the overall picture of the evidence. They go beyond individual study results and explore patterns, consistencies, and discrepancies across studies. Through synthesis, systematic reviews provide a broader understanding of the research question.

Identification of Knowledge Gaps

Systematic reviews often reveal gaps in the existing literature. By identifying areas where evidence is limited or conflicting, they highlight the need for further research and guide future study design and research priorities.

Meta-Analysis

Meta-analysis is a statistical technique used to combine and analyze data from multiple independent studies addressing the same research question. It aims to provide a more precise estimate of the treatment effect or association by pooling data from individual studies, increasing the overall sample size, and enhancing the statistical power. Meta-analysis is a critical component of systematic reviews, which are comprehensive reviews of the literature on a specific topic that follow a predefined methodology. Here are some reasons why meta-analysis are a valuable and resourceful tool.

1. **Increased Precision:** Meta-analysis combines data from multiple studies, increasing the sample size and statistical power. This larger pool of participants enhances the precision of the estimated treatment effect, providing a more accurate representation of the intervention's impact.
2. **Detection of Small Treatment Effects:** Meta-analysis can detect small treatment effects that may not be evident in individual studies due to limited sample sizes. By aggregating data across studies, meta-analysis enhances the ability to detect subtle differences between treatment groups.
3. **Assessment of Treatment Consistency:** Meta-analysis enables the examination of treatment effects across different populations, settings, and study designs. It helps identify the consistency or heterogeneity of treatment effects, providing insights into factors that may influence intervention outcomes.
4. **Subgroup Analysis:** Meta-analysis allows for subgroup analyses to explore potential effect modifiers. By examining specific subgroups, such as age, gender, or disease severity, researchers can determine if the treatment effect varies across different patient characteristics.

Limitations of Systematic Reviews and Meta-Analyses

1. **Quality of Included Studies:** The quality of systematic reviews and meta-analyses depends on the quality of the studies included. If the primary studies are of poor quality or have biases, the findings of the systematic review may be limited or compromised.
2. **Publication Bias:** Systematic reviews are susceptible to publication bias, where studies with statistically significant results are more likely to be published, leading to an overestimation of treatment effects. However, efforts are made to identify and include unpublished studies to mitigate this bias.
3. **Heterogeneity:** Variability across studies, known as heterogeneity, can pose challenges in meta-analysis. Differences in study design, populations, interventions, and outcome measures may contribute to heterogeneity. Statistical methods, such as subgroup analysis or sensitivity analysis, are employed to explore and address heterogeneity.

Conclusion

In summary, systematic reviews and meta-analyses are vital tools in clinical research, providing a comprehensive and evidence-based summary of existing literature. They contribute to the understanding of treatment efficacy, inform clinical guidelines, and guide evidence-based decision-making in healthcare. However, it is essential to interpret the findings of systematic reviews and meta-analyses in conjunction with other sources of evidence, clinical expertise, and patient preferences to make well-informed and individualized decisions for patient care.

SECTION 6: MIXED METHODS RESEARCH

Mixed Methods Research

In this section, we will discuss mixed methods research, which combines quantitative and qualitative approaches. We will explain the rationale behind mixed methods research, which allows researchers to gain a comprehensive understanding of complex phenomena by integrating numerical data with qualitative insights. We will discuss the design and analysis considerations involved in mixed methods research, emphasizing its potential to generate holistic and context-specific knowledge. Readers will gain an understanding of how mixed methods research contributes to a more comprehensive understanding of healthcare-related issues.

Introduction

Mixed methods research is an approach that combines quantitative and qualitative research methods to gain a comprehensive understanding of complex phenomena. It recognizes that using both numerical data and qualitative insights can provide a more holistic view of the research topic, enriching the understanding of the context, processes, and meanings associated with the phenomenon under investigation.

The rationale behind mixed methods research is rooted in the belief that no single method can fully capture the complexity of many research questions. By integrating quantitative and qualitative data, researchers can overcome the limitations of each method and gain a more nuanced understanding of the research topic. Here are key aspects to consider:

Design Considerations

- **Sequential Design:** In a sequential design, the quantitative and qualitative components are conducted in a sequence, with one method informing the other. For example, a study may begin with a quantitative survey to establish prevalence rates, followed by qualitative interviews to explore participants' experiences.
- **Concurrent Design:** In a concurrent design, both quantitative and qualitative data are collected simultaneously. The data are then analyzed independently, and the findings are integrated during the interpretation phase. This design allows for the exploration of convergence or divergence between the two data sources.
- **Embedded Design:** An embedded design involves one method being subordinate to the other, with one method providing a supportive role to enhance or enrich the other. For example, a survey may be supplemented with qualitative interviews to gain a deeper understanding of participants' responses.

Data Collection

- **Quantitative Data:** Quantitative data collection methods commonly used in mixed methods research include surveys, experiments, and structured observations. These methods generate numerical data that can be analyzed using statistical techniques to identify patterns, correlations, and statistical significance.
- **Qualitative Data:** Qualitative data collection methods include interviews, focus groups, and observations. These methods aim to explore participants' perspectives, experiences, and meanings. Qualitative data are typically collected through open-ended questions, narrative accounts, or observations.

Data Analysis

- **Quantitative Analysis:** Quantitative data are analyzed using statistical methods to describe, compare, and identify relationships or associations between variables. Common techniques include descriptive statistics, inferential statistics, and multivariate analysis.
- **Qualitative Analysis:** Qualitative data analysis involves organizing, coding, and interpreting the data to identify themes, patterns, and categories. Techniques such as thematic analysis, content analysis, or grounded theory can be employed to explore the richness and depth of qualitative data.

Integration of Findings

- **Convergence:** Convergence occurs when quantitative and qualitative findings support and reinforce each other. The convergence of findings adds depth and context to quantitative results, helping to explain the mechanisms or processes underlying the observed patterns.
- **Expansion:** Expansion occurs when quantitative and qualitative findings provide complementary but distinct insights into the research topic. The qualitative data may reveal new dimensions or perspectives not captured by quantitative measures, expanding the understanding of the phenomenon.
- DIVERGENCE: Divergence occurs when quantitative and qualitative findings present contrasting or conflicting perspectives. These discrepancies can be explored to uncover the reasons behind the discrepancies and gain a more nuanced understanding of the research topic.

The strengths of mixed methods research lie in its ability to generate a comprehensive understanding of complex phenomena by integrating numerical data with qualitative insights. By combining different types of data, researchers can address research questions from multiple angles, capture the breadth and depth of the topic, and provide a more contextualized understanding. Mixed methods research is particularly valuable in healthcare-related research, where understanding the interplay of factors, patient experiences, and social contexts is crucial for informing practice and policy.

However, it's important to consider the challenges and limitations of mixed methods research. Integration of different data types requires careful planning and expertise, and the analysis process can be time-consuming. Additionally, mixed methods research may require a larger sample size and resources to accommodate both quantitative and qualitative components. Ensuring rigor and quality in data collection, analysis, and interpretation is paramount to maintain the validity and trustworthiness of the findings.

Conclusion

In conclusion, mixed methods research offers a powerful approach to address complex research questions in healthcare. By integrating quantitative and qualitative methods, it enhances the depth and breadth of understanding, providing a more comprehensive and contextualized view of the phenomena under study. The choice of mixed methods research should be guided by the research question and the unique characteristics of the research topic, allowing for a more nuanced exploration and interpretation of the data.

4

DESIGNING AND PLANNING A CLINICAL RESEARCH STUDY

CHAPTER IV

DESIGNING AND PLANNING A CLINICAL RESEARCH STUDY

Effective design and meticulous planning are the cornerstones of successful clinical research. This chapter provides an in-depth exploration of the various methodologies involved in designing and planning clinical research studies, with a focus on ensuring scientific integrity, relevance, and ethical compliance. By understanding the fundamental principles and steps in the study design and planning process, researchers can create robust studies that yield reliable, valid, and actionable insights.

SECTION 1: FORMULATING A RESEARCH QUESTION

Formulating a Research Question

The foundation of any successful clinical research study is a well-formulated research question. This section outlines the process of developing a clear, concise, and effective research question, which is essential for guiding the design, methodology, and direction of the study. A strong research question not only specifies what the researcher aims to answer but also influences the choice of study design and data analysis approach.

The Importance of a Well-Formulated Research Question

Formulating a well defined research question is a critical step in the process of conducting a clinical research study. The research question serves as the foundation of the study, guiding its objectives, study design, and data collection. It plays a crucial role in identifying knowledge gaps, addressing clinical observations, and incorporating emerging scientific advancements.

Characteristics of a Good Research Question

A good research question possesses several key characteristics. First, it should be specific, clearly defining the scope of the research inquiry. Second, it should be measurable, identifying observable and quantifiable outcomes that can be assessed. Third, the research question should be achievable within the available resources and time frame of the study. Fourth, it should be relevant, addressing a significant issue or gap in knowledge within the field. Finally, the research question should be time-bound, specifying a clear timeline for data collection and analysis, following the SMART criteria.

- **Specificity:** Clearly Defining the Scope of the Research Question
- **Measurability:** Identifying Observable and Quantifiable Outcomes
- **Achievability:** Ensuring Feasibility within the Available Resources and Timeframe
- **Relevance:** Addressing a Significant Issue or Gap in Knowledge
- **Time-Bound:** Specifying a Clear Timeline for Data Collection and Analysis (SMART Criteria)

Research Question Formulation Process

The process of formulating a research question involves several steps:

- **Literature Review:** Conducting a thorough review of existing literature to identify what has been done, highlight gaps, and refine the area of interest.
- **Identifying Gaps:** Pinpointing specific gaps within the literature where new data could significantly impact current understanding or practice.
- **Drafting Preliminary Questions:** Creating a list of

potential questions and refining them based on feedback from peers or mentors in the field.

- **Utilizing Frameworks:** Applying frameworks such as PICO (Population, Intervention, Comparator, Outcome) to structure the question effectively and ensure all components are clearly defined.
- **Final Selection:** Choosing a question that best meets the criteria for specificity, relevance, feasibility, and complexity, ensuring it aligns with the goals of the research and the needs of the target population.

Refining the Research Question

Refining the research question is an iterative process. Researchers consider ethical considerations and participant safety in the research design. Seeking input from peers, mentors, and research advisory committees helps ensure clarity and relevance. Piloting and pretesting the research question are essential steps to validate and refine the question.

Example of Research Question Development

To illustrate, consider a researcher interested in diabetes management. After a literature review, the researcher identifies a gap in the effectiveness of digital health interventions for managing Type 2 Diabetes. A preliminary question might be, "Do digital health interventions improve blood sugar control in Type 2 Diabetic patients compared to traditional management methods?" This question can then be refined with specifics about the type of digital interventions, the patient population, and the metrics for measuring blood sugar control.

Conclusion

A well-formulated research question is crucial in clinical research as it guides the entire study process. By understanding the characteristics of a good research question and the iterative process of refining it, researchers can conduct studies that advance medical knowledge and improve patient care. The formulation of a research question ensures that the study objectives are clear, the design is appropriate, and the outcomes are relevant to addressing gaps in scientific understanding.

SECTION 2: STUDY DESIGN SELECTION

Study Design Selection

Once a research question has been formulated, selecting the most suitable study design is the next critical step. The choice of study design is pivotal because it directly affects the accuracy and applicability of the results. This section outlines the key considerations in selecting a study design and describes the main types of designs used in clinical research.

Introduction

The study design is the blueprint of a clinical research project, defining its framework and guiding the methodology to ensure reliable and valid results. It includes various types of designs, such as randomized controlled trials (RCTs), cohort studies, case-control studies, cross-sectional studies, and qualitative research. Each design type has its own strengths and limitations, making it crucial for researchers to select the most appropriate one based on the research question and objectives.

For instance, RCTs are considered the gold standard in clinical research due to their ability to minimize bias through randomization and blinding. A practical example is the RCT conducted to evaluate the effectiveness of the COVID-19 vaccines, where participants were randomly assigned to receive either the vaccine or a placebo. This design ensured that the results were not influenced by participant or researcher bias and provided robust evidence on the vaccine's efficacy and safety.

Key Considerations for Selecting a Study Design

Choosing the right study design involves several important considerations:

- **Objectives of the Study:** Different designs are better suited to different types of research questions. For instance, questions about efficacy might best be answered with a randomized controlled trial, while questions about prevalence might require a cross-sectional study.
- **Resources Available:** Some study designs require more resources than others. Considerations include funding, available expertise, time constraints, and available participants.
- **Ethical Implications:** Certain designs might pose greater ethical concerns, particularly those involving vulnerable populations or potentially risky interventions.
- **Statistical Considerations:** The ability to gather enough data to achieve meaningful results without overburdening resources is crucial. The chosen design should facilitate a clear statistical analysis plan that can effectively address the research question.

Overview of Study Designs

The following are some of the primary types of study designs used in clinical research, each suited for different kinds of research questions:

Experimental Designs

Experimental designs, such as randomized controlled trials (RCTs) are best for testing the efficacy and safety of a new intervention. Participants are randomly assigned to either the treatment group or the control group to measure the effects of the intervention.

Observational Designs

Observational studies are commonly used when the research question involves observing and collecting data on participants in their natural settings. They do not involve any intervention or manipulation of variables. Three main types of observational studies include:

- **Cohort Studies:** Suitable for examining the effects of certain exposures over time, cohort studies follow a group of people who share a common characteristic or experience within a defined period.
- **Case-Control Studies:** Often used in situations where the outcome is rare, this design involves comparing participants with a condition (cases) to those without (controls) to identify factors that might contribute to the condition.
- **Cross-Sectional Studies:** Useful for assessing the prevalence of an outcome or condition at a single point in time, providing a snapshot of a situation.

Qualitative Designs

These designs are geared toward exploring broad thematic questions, often using interviews, focus groups, and observations to collect data that are descriptive rather than quantitative.

Case Example

Imagine a study aiming to determine the impact of a dietary intervention on heart disease risk. Given the direct health implications, an experimental design like a randomized controlled trial might be ideal for accurately assessing the effects of the diet. However, if the intervention poses potential risks, an observational cohort study might be considered to first observe the outcomes of natural dietary patterns on heart disease over time without manipulating the diet of the participants.

Conclusion

By the end of this section, readers should have a solid understanding of how to align their research question with an appropriate study design, considering the study's goals, available resources, ethical considerations, and required analytical power. This alignment is crucial for the successful execution and validity of clinical research findings.

SECTION 3: DEVELOPING THE RESEARCH PROTOCOL

Developing the Research Protocol

After choosing a suitable study design, the next crucial step in planning a clinical research study is the development of a comprehensive research protocol. This document serves as a roadmap for conducting the study, ensuring all involved parties maintain rigorous adherence to the agreed-upon procedures and ethical standards. This section outlines the essential components of a research protocol and the process of creating an effective document.

Introduction

The development of a research protocol is a critical step that outlines the study's objectives, design, methodology, and procedures in detail. A well-structured protocol ensures consistency, reproducibility, and ethical compliance throughout the study. It includes sections such as the study objectives, study design, participant selection criteria, intervention details, data collection methods, and statistical analysis plan.

For example, a clinical trial investigating a new diabetes medication would include detailed protocols on dosage, administration schedule, monitoring of blood glucose levels, and criteria for participant inclusion and exclusion. By adhering to the protocol, researchers can ensure that the study is conducted systematically and ethically, thereby enhancing the credibility of the findings.

Importance of the Research Protocol

A research protocol is vital as it details the objective(s), design, methodology, statistical considerations, and organization of a study. It ensures that the study adheres to regulatory requirements, ethical guidelines, and scientific standards. The protocol also provides a point of reference for all study-related activities and is crucial for gaining approval from regulatory bodies and ethics committees.

Components of a Research Protocol

An effective research protocol typically includes the following components:

1. **Background and Rationale:** This section provides the scientific background, previous studies on the topic, and the rationale behind the research question and chosen methodology.
2. **Objectives:** Clearly defined primary and secondary objectives of the study based on the research question.
3. **Study Design:** Detailed description of the study design chosen, explaining the type of study, procedures involved, and the rationale for the design choice.
4. **Subject Selection and Exclusion Criteria:** Criteria for participant inclusion and exclusion, ensuring the selected participants are appropriate for addressing the research question.

5. **Interventions:** If applicable, a detailed description of the interventions to be used, including drugs, devices, procedures, and control conditions.
6. **Outcome Measures:** Definition of primary and secondary endpoints, including how and when they will be measured.
7. **Sample Size Calculation:** Explanation of the statistical methods used to determine the number of participants needed to achieve adequate power.
8. **Data Collection Methods:** Procedures for data collection, management, and analysis plans, ensuring data integrity and confidentiality.
9. **Ethical Considerations:** Discussion of how ethical standards will be maintained, including informed consent processes and measures to protect participant privacy and data.
10. **Timeline:** Estimated durations for each phase of the study, from initiation through to data analysis and report writing.
11. **Budget and Funding:** Detailed budget, funding sources, and justification for the costs involved.

Developing the Protocol

Developing a research protocol involves several key steps:

1. **Literature Review and Preliminary Studies:** Conduct thorough research to ensure the study is necessary and not duplicative, and that it builds logically on previous work.
2. **Consultation with Stakeholders:** Engage with various stakeholders, including other researchers, healthcare professionals, and patient advocacy groups, to refine the study design and protocol.
3. **Drafting and Revising:** Write the initial draft of the protocol and revise it based on feedback from these consultations.
4. **Ethical Approval:** Submit the protocol to an institutional review board (IRB) or ethics committee for review and approval, addressing any concerns they raise.

Case Example

Consider a protocol for a study investigating the effect of a new cardiac rehabilitation program on post-myocardial infarction recovery. The protocol would detail the theoretical basis for the intervention, describe the randomized controlled trial design, specify the exercise and education components of the rehabilitation program, outline the criteria for patient inclusion, and define the primary outcome measures, such as improvements in cardiovascular health metrics over a specified amount of time.

Conclusion

By the end of this section, readers should understand how to methodically develop a research protocol that is both comprehensive and compliant with ethical and regulatory standards. This document is crucial for the systematic execution of the research and upholding the integrity of the study's findings.

SECTION 4: ETHICAL CONSIDERATIONS IN STUDY DESIGN

Ethical Considerations in Study Design

Ethical integrity is paramount in clinical research, influencing every aspect from the formulation of the research question to the dissemination of the findings. This section outlines the ethical considerations that must be integrated into the design and planning of clinical studies to ensure the protection of participants and the credibility of the research results.

Ethical Frameworks in Clinical Research

Before getting into specific ethical considerations, it is crucial to understand the frameworks that guide ethical decisions in research. These include the Belmont Report, the Declaration of Helsinki, and the Common Rule.

- **The Belmont Report:** Emphasizes respect for persons, beneficence, and justice.
- **Declaration of Helsinki:** Focuses on the ethical principles for medical research involving human subjects, including considerations for post-study access.
- **Common Rule:** Governs the ethical conduct and requirements for informed consent, IRB reviews, and assurance of compliance.

Incorporating Ethical Considerations in Research Design

The ethical design of a clinical study is not a one-time consideration but a continuous commitment throughout the research lifecycle. Key areas to address include:

- **Participant Safety and Well-being:** Safety is the most critical concern. Researchers must design studies to minimize potential risks and discomforts to participants. This involves choosing study methods that are proportional to the potential benefits and ensuring that any risk is justifiable by the anticipated outcomes.
- **Informed Consent:** One of the most crucial components of ethical research is ensuring informed consent. It is essential to design consent forms and processes that are easily understandable, transparent, and culturally

sensitive. Participants must be fully informed about the purpose, procedures, risks, benefits, and their rights to withdraw from the study without any consequences.

- **Privacy and Confidentiality:** Protecting the privacy and confidentiality of participant data is fundamental. Researchers must implement secure methods for handling and storing sensitive information and share data only as necessary, under strict confidentiality agreements.

- **Equity and Inclusion:** Ethical research design also involves ensuring that the study sample appropriately reflects the diversity of the population that the research is intended to benefit. This includes considering age, gender, ethnicity, socio-economic status, and other demographics to avoid biases and enhance the generalizability of the study findings.

- **Vulnerable Populations:** Special considerations are needed when the study involves vulnerable populations such as children, pregnant women, or individuals with cognitive impairments. Studies must include additional safeguards to protect these groups from exploitation or harm.

Ethical Approval Process

Obtaining ethical approval is a prerequisite for conducting any clinical research involving human subjects. This process ensures that the study complies with both international and national ethical guidelines, protecting participants throughout the research. Below is a detailed examination of the steps involved in the ethical approval process and strategies for successful navigation.

Understanding the Role of Institutional Review Boards (IRBs)

The primary role of an Institutional Review Board (IRB), or equivalent ethics committee, is to review research proposals to ensure that they meet ethical standards. The IRB's responsibilities include:

- **Review of the research protocol:** The IRB examines the study's aims, methods, and materials to ensure that they are ethically sound and that the benefits justify the risks.

- **Assessment of participant safety measures:** The IRB evaluates procedures for protecting participants from potential harms, including physical, psychological, and privacy risks.

- **Evaluation of informed consent process:** This involves reviewing the consent forms and procedures to ensure that they are clear, comprehensive, and capable of being easily understood by participants.

Preparing for IRB Submission

Preparation is key to a smooth IRB review process. The following steps are crucial:

1. **Develop a detailed protocol:** Include all aspects of the study, such as background, objectives, methodology, participant recruitment strategies, data management, and analysis plans.
2. **Prepare informed consent documents:** These should be clear and include all information required by ethical guidelines, such as study risks, benefits, alternatives to participation, and participants' rights.
3. **Risk assessment documentation:** Clearly outline potential risks and their management strategies to demonstrate proactive safety measures.

The Review Process

The review process typically involves the following stages:

1. **Initial review:** The IRB conducts a preliminary review of the submission to check for completeness and any immediate ethical concerns.
2. **Full review:** For studies involving more than minimal risk, a full committee review is usually required. This often involves a meeting where researchers may be asked to present their study and respond to questions.
3. **Expedited review:** For minimal-risk research, an expedited review may be conducted by one or more experienced reviewers from the IRB without convening a full committee.
4. Conditional approval or modifications required: Often, the IRB may approve a protocol conditionally, requesting modifications or additional information before final approval.
5. **Approval:** Once the IRB is satisfied that the ethical concerns have been adequately addressed, formal approval is granted.

Post-Approval Monitoring

Ethical oversight does not end with IRB approval. Post-approval monitoring is conducted to ensure ongoing compliance with the approved protocol and ethical standards. This may include:

- **Continuing review:** Regular reviews by the IRB, especially for long-term studies. Most IRB's will require a Continuing Review and a trial status update annually.

- **Reporting of adverse events:** Immediate reporting to the IRB of any unexpected complications or risks encountered during the study.

- **Amendments to the study:** Any changes to the study protocol or procedures must be approved by the IRB before implementation. This includes changes to the study population (such as an update to an age range or patients from a specific geographical location), wording on a consent form, or changes to study personnel.

Tips for Facilitating Ethical Approval

- **Engage with the IRB early:** Early discussions can help clarify expectations and improve the quality of your submission.

- **Ensure clarity and thoroughness in all documentation:** Clear, detailed submissions can facilitate a smoother review process.

- **Be responsive to IRB feedback:** Address any concerns or requests for modifications promptly and thoroughly.

Conclusion

This section aims to equip researchers with the knowledge and tools needed to integrate ethical considerations into their study designs systematically, ensuring research integrity and respect for all participants involved. By understanding and applying these ethical principles, researchers can not only comply with regulatory standards but also foster trust and transparency in clinical research.

SECTION 5: PRACTICAL ASPECTS OF CONDUCTING CLINICAL RESEARCH

Practical Aspects of Conducting Clinical Research

The success of a clinical study depends not only on rigorous scientific and ethical foundations but also on effective management of practical aspects. This section covers essential operational components including budgeting, resource management, participant recruitment, and retention, providing strategies for effectively navigating these challenges.

Budgeting and Funding

Adequate funding and meticulous budgeting are crucial for the smooth execution of a clinical study. In this section, we will discuss some key considerations for effective financial planning.

- **Estimating Costs:** Begin with a detailed breakdown of all possible costs, including personnel salaries, equipment, supplies, participant compensation, and indirect costs such as administrative support and facility usage fees.

- **Securing Funding:** Explore various funding sources such as government grants, institutional funding, and private sector sponsorship. Be prepared to justify the budget in funding applications clearly and comprehensively.

- **Financial Management:** Implement rigorous accounting procedures to track expenses and ensure the project remains within budget. Regular financial reviews can help identify and rectify overspending or reallocate funds as needed.

Resource Allocation

Efficient use of resources, including human resources, equipment, and facilities are essential for conducting research effectively.

- **Human Resources:** Allocate roles and responsibilities clearly among team members to optimize the use of expertise and labor. Regular training and updates can help maintain high performance and adapt to evolving project needs.

- **Equipment and Facilities:** Ensure that all necessary equipment and facilities are available and in good working condition. Plan for maintenance and potential upgrades to avoid disruptions in the study timeline.

- **Time Management:** Develop a detailed project timeline with key milestones and deadlines. Regular progress checks can help keep the project on track and adjust timelines as necessary.

Participant Recruitment and Retention

Recruiting and retaining participants are among the most challenging aspects of clinical research and can be costly if not thought through beforehand.

- **Recruitment Strategies:** Develop a comprehensive recruitment plan that includes identification of target populations, use of recruitment channels (e.g., clinics, community centers, social media), and clear, engaging recruitment materials.

- **Inclusion and Diversity:** Aim for a diverse participant pool to enhance the generalizability of the study results. Address potential barriers to participation, such as language, transportation, or childcare.

- **Retention Efforts:** Implement strategies to minimize dropout rates, such as regular follow-up, participant incentives, and addressing participants' concerns and needs promptly.

Managing Data Collection and Quality Assurance

Effective data management is critical to ensure the accuracy and reliability of research findings.

- **Data Collection Systems:** Utilize reliable and validated tools for data collection. Electronic data capture systems can enhance efficiency and reduce errors.

- **Quality Assurance:** Establish quality control procedures to check data accuracy and completeness regularly. Training for data collectors on the importance of fidelity to data collection protocols is crucial.

- **Data Security:** Ensure compliance with data protection regulations. Implement secure data storage and transfer methods to protect participant confidentiality.

Navigating Regulatory Requirements

Understanding and adhering to regulatory requirements is essential for clinical research and is necessary to understand for successful clinical studies.

- **Regulatory Oversight:** Stay informed about the regulatory landscape, which can vary by location and type of research. Prepare for inspections and audits by regulatory bodies.

- **Documentation and Reporting:** Maintain comprehensive records of all study activities as required by regulatory authorities. Prepare for potential audits by ensuring documentation is complete, up-to-date, and readily accessible.

Conclusion

By effectively managing these practical aspects, researchers can enhance the efficiency, impact, and ethical compliance of their clinical studies. This section aims to provide foundational strategies that are essential for navigating the complex logistics of conducting clinical research.

SECTION 6: DATA COLLECTION AND MEASUREMENT TOOLS

Data Collection and Measurement Tools

Data collection is a critical component of clinical research that involves gathering information to address research questions and test hypotheses. This section aims to provide a thorough understanding of the data collection process and the selection of appropriate measurement tools. It will highlight the importance of using reliable and valid instruments to ensure accurate and meaningful data. Various data collection methods, such as interviews, surveys, observations, medical records review, and laboratory tests, will be discussed. Considerations for selecting or developing measurement tools, including psychometric properties, feasibility, cultural appropriateness, and previous use in similar studies, will also be explored. Additionally, guidance on data collection techniques and best practices to ensure data quality and integrity will be provided.

Introduction

Data collection is a fundamental aspect of clinical research, involving the systematic gathering of information relevant to the research objectives. Accurate and reliable data collection is essential for drawing valid conclusions. Methods of data collection vary depending on the study design and can include surveys, interviews, laboratory tests, imaging techniques, and electronic health records.

A case study illustrating effective data collection is the Framingham Heart Study, which has been ongoing since 1948. This longitudinal study collects comprehensive data on cardiovascular health from participants over several decades. The extensive data collected through physical examinations, laboratory tests, and questionnaires have led to significant insights into the risk factors for heart disease, highlighting the importance of meticulous data collection in clinical research.

Importance of Reliable and Valid Measurement Tools

Using reliable and valid measurement tools is crucial in clinical research to ensure the accuracy and meaningfulness of the collected data. Reliability refers to the consistency and stability of the instrument, while validity refers to the extent to which the instrument measures what it is intended to measure. By using reliable and valid instruments, researchers can obtain trustworthy and credible data that can be confidently interpreted and used to draw conclusions.

Data Collection Methods

Interviews: Interviews involve direct interaction with participants to gather information. They can be structured (using standardized questions) or unstructured (allowing for open-ended responses). Interviews provide in-depth insights into participants' experiences, perceptions, and attitudes. Researchers must establish rapport, use appropriate questioning techniques, and document responses accurately.

Surveys: Surveys utilize standardized questionnaires or forms to collect data from a larger sample. Surveys can be self-administered (paper-based or online) or interviewer-administered. They allow for efficient data collection but require careful questionnaire design to ensure clarity, comprehensiveness, and relevance to the research objectives.

Observations: Observational methods involve systematically watching and recording behaviors, interactions, or events. Observations can be structured (following a predefined checklist) or unstructured (allowing for more flexibility). This method provides firsthand information on participants' behaviors in natural or controlled settings. It requires well-trained observers, detailed protocols, and appropriate ethical considerations.

Medical Records Review: Reviewing medical records allows researchers to collect data on participants' medical history, diagnoses, treatments, and outcomes. This method relies on existing clinical documentation and can provide valuable information for retrospective studies. Ethical considerations, data privacy, and data extraction protocols need to be carefully addressed.

Laboratory Tests: Laboratory tests involve the collection and analysis of biological samples to obtain objective data on various physiological or biochemical parameters. These tests provide quantitative data and can be used to evaluate biomarkers, drug concentrations, genetic profiles, or other relevant measures. Standardized protocols, quality control measures, and appropriate laboratory facilities are essential.

Selection and Development of Measurement Tools

When selecting or developing measurement tools, several factors should be considered:

Psychometric Properties: Instruments should have established reliability and validity. Psychometric properties, such as test-retest reliability, internal consistency, construct validity, and criterion validity, should be evaluated. Existing instruments with established psychometric properties are preferred, but when suitable instruments are unavailable, researchers may need to develop their own and conduct psychometric testing.

Feasibility: Consider the resources, time, and skills required to administer and analyze data using a particular tool. Feasibility includes aspects such as cost-effectiveness, ease of administration, and the ability to reach the target population.

Cultural Appropriateness: Ensure that the measurement tool is culturally appropriate, considering language, cultural norms, and values of the study population. Translations and adaptations may be necessary to ensure the tool's relevance and comprehensibility.

Previous Use: Consider whether the measurement tool has been used in previous studies with similar objectives or populations. Utilizing established tools with a documented track record strengthens the comparability of results across studies.

Data Collection Techniques and Best Practices

To ensure data quality and integrity, researchers should consider the following best practices:

Training and Standardization: Ensure that data collectors are appropriately trained and adhere to standardized data collection protocols. This minimizes variability and enhances consistency in data collection.

Pilot Testing: Conduct a pilot study to assess the feasibility and validity of the data collection procedures. This allows for refinement of the measurement tools and identification of any potential issues before full-scale implementation.

Data Monitoring and Quality Control: Implement mechanisms for data monitoring, quality control, and data cleaning. Regular checks and validation procedures should be in place to identify and address data entry errors, missing data, and outliers.

Confidentiality and Data Security: Safeguard participant confidentiality and data security throughout the data collection process. Establish protocols for data storage, access, and retention that comply with relevant ethical and legal requirements.

Conclusion

Data collection is a crucial aspect of clinical research, and selecting appropriate measurement tools is essential to ensure the accuracy and meaningfulness of collected data. Using reliable and valid instruments enhances the trustworthiness of the findings and supports the validity of conclusions. Researchers should carefully consider the various data collection methods, such as interviews, surveys, observations, medical records review, and laboratory tests, based on the research objectives and target population. When selecting or developing measurement tools, psychometric properties, feasibility, cultural appropriateness, and previous use in similar studies should be considered. Adhering to best practices and implementing quality control measures throughout the data collection process ensures data integrity and enhances the overall validity of the study.

SECTION 7: SAMPLE SIZE CALCULATION

Sample Size Calculation

Determining an appropriate sample size is a critical aspect of clinical research as it directly affects the study's statistical power and the ability to detect meaningful effects or differences between groups. This section aims to provide a thorough understanding of the importance of sample size determination and the factors that influence it. We will also discuss common methods for sample size estimation, including power calculations for hypothesis testing and considerations for qualitative research.

Importance of Sample Size

Having an adequate sample size is crucial for obtaining reliable and valid results in clinical research. Insufficient sample sizes may lead to underpowered studies, where the statistical power to detect true effects is low. This can result in Type II errors, failing to detect significant differences or effects when they actually exist. By determining an appropriate sample size, researchers can enhance the statistical power of their study, increasing the chances of detecting meaningful effects or differences.

Factors Influencing Sample Size

Several factors influence the determination of sample size. The desired level of significance (alpha), which denotes the probability of rejecting the null hypothesis when it is true, is typically set at 0.05. Affect size, or the magnitude of the effect being investigated, directly affects sample size requirements. Larger effect sizes require smaller sample sizes to detect significant differences. The variability of the outcome under investigation, measured by the standard deviation, also plays a role. Greater variability necessitates larger sample sizes to achieve adequate power. Additionally, the study design, such as whether it is a randomized controlled trial or an observational study, influences the sample size calculation.

Methods for Sample Size Estimation

Power calculations are commonly used to estimate the sample size needed for hypothesis testing in quantitative research. These calculations consider factors like the desired level of significance, effect size, variability, and power (typically set at 0.80 or 0.90). Various statistical methods, such as formulas or software packages, are available to conduct power calculations based on the specific study design and statistical tests being employed. Researchers can consult resources like sample size calculators or statistical textbooks to guide them through this process.

In qualitative research, sample size determination is often based on the concept of data saturation. Data saturation occurs when new data no longer provide additional insights or themes, indicating that theoretical saturation has been reached. Researchers typically aim to interview or observe enough participants to achieve data saturation, ensuring that they have captured a comprehensive understanding of the phenomenon under investigation. While there are no fixed sample size formulas in qualitative research, researchers can draw on published guidelines and previous studies as references to inform their decision-making.

Practical Examples and Resources

To illustrate the concept of sample size determination, let's consider an example. Suppose a researcher is conducting a clinical trial to evaluate the effectiveness of a new intervention on reducing pain in patients with a specific condition. Based on previous studies and expert knowledge, the researcher estimates the effect size to be moderate. Using a power calculation with a desired power of 0.80, a desired level of significance of 0.05, and the estimated effect size and variability, the researcher determines that a sample size of 100 participants is necessary to detect a significant difference between groups.

Researchers can find a variety of resources to guide them in calculating sample sizes for their own studies. Statistical software packages like R, SPSS, or G*Power provide tools for power calculations in quantitative research. Additionally, textbooks on research methods, biostatistics, or clinical trial design often contain detailed explanations and step-by-step guides for sample size determination.

Conclusion

Determining an appropriate sample size is crucial for ensuring the statistical power and validity of clinical research studies. By considering factors such as the desired level of significance, effect size, variability, and study design, researchers can estimate the sample size needed to detect meaningful effects or differences between groups. Power calculations are commonly used in quantitative research, while qualitative research relies on achieving data saturation. Researchers can consult various resources, including software packages and textbooks, to assist them in calculating sample sizes for their own studies.

SECTION 8: STATISTICAL ANALYSIS AND DATA INTERPRETATION

Statistical Analysis and Data Interpretation

The statistical analysis and data interpretation phase is a crucial component of clinical research that involves analyzing collected data to draw valid conclusions and make meaningful inferences. This section aims to provide a thorough understanding of the importance of employing appropriate statistical methods, common statistical techniques, and the interpretation of statistical results. It will cover descriptive statistics, inferential statistics (e.g., t-tests, chi-square tests), regression analysis, survival analysis, and qualitative data analysis approaches (e.g., thematic analysis, grounded theory). Additionally, it will emphasize the need to consider clinical significance alongside statistical significance and the importance of transparent reporting of statistical methods and results.

Importance of Employing Appropriate Statistical Methods

Employing appropriate statistical methods is crucial in clinical research to ensure the accuracy and validity of the conclusions drawn from the collected data. Statistical analysis allows researchers to summarize, explore, and make inferences about the data. It helps to identify patterns, relationships, and associations, and enables comparisons between groups or variables of interest. By using appropriate statistical methods, researchers can address research questions, test hypotheses, and make evidence-based decisions.

Common Statistical Techniques

Descriptive Statistics: Descriptive statistics summarize and describe the characteristics of the collected data. Measures such as mean, median, standard deviation, and percentages are used to provide a concise summary of the data distribution and central tendency. Descriptive statistics help to summarize data in a meaningful and interpretable way.

Inferential Statistics: Inferential statistics involve making inferences about a population based on a sample. Common techniques include t-tests, chi-square tests, analysis of variance (ANOVA), and correlation analysis. These techniques allow researchers to test hypotheses, determine significant differences between groups, and examine associations between variables.

Regression Analysis: Regression analysis is used to examine the relationship between a dependent variable and one or more independent variables. It helps to assess the impact of predictor variables on the outcome variable and quantify their associations. Linear regression, logistic regression, and Cox regression (for survival analysis) are commonly used regression techniques in clinical research.

Survival Analysis: Survival analysis is used when studying time-to-event outcomes, such as the time to disease progression or death. It allows researchers to estimate survival probabilities over time and examine the impact of various factors on survival outcomes. Techniques such as Kaplan-Meier analysis and Cox proportional hazards regression are commonly employed in survival analysis.

Qualitative Data Analysis: Qualitative data analysis involves analyzing textual or narrative data, such as interview transcripts or open-ended survey responses. Approaches like thematic analysis and grounded theory are commonly used to identify patterns, themes, and categories within the qualitative data. These techniques help to provide rich insights into participants' experiences, perceptions, and perspectives.

Interpretation of Statistical Results

Interpreting statistical results involves making sense of the findings in the context of the research objectives and drawing meaningful conclusions. It is important to consider both statistical significance and clinical significance. Statistical significance indicates whether an observed effect is likely to have occurred by chance, whereas clinical significance pertains to the practical or meaningful impact of the findings on patient outcomes or healthcare practice. Researchers should carefully interpret the effect sizes, confidence intervals, p-values, and consider the clinical relevance of the results.

Transparent Reporting of Statistical Methods and Results

Transparent reporting of statistical methods and results is essential to ensure research reproducibility and facilitate critical appraisal. Researchers should provide clear descriptions of the statistical techniques employed, including assumptions, software used, and any adjustments made for confounding factors. It is important to report effect sizes, measures of variability, p-values, and confidence intervals to accurately communicate the findings. Adhering to reporting guidelines, such as those provided by the Consolidated Standards of Reporting Trials (CONSORT) or the Strengthening the Reporting of Observational Studies in Epidemiology (STROBE), enhances transparency and enables readers to assess the validity and reliability of the study.

Conclusion

The statistical analysis and data interpretation phase of a clinical research study is vital for drawing valid conclusions and making evidence-based decisions. Employing appropriate statistical methods allows researchers to analyze data, explore relationships, and test hypotheses. Descriptive and inferential statistics, regression analysis, survival analysis, and qualitative data analysis approaches are commonly employed in clinical research. Interpreting statistical results involves considering both statistical significance and clinical significance. Transparent reporting of statistical methods and results ensures research reproducibility and facilitates critical appraisal. By understanding and appropriately applying statistical techniques, researchers can derive meaningful insights from the collected data and contribute to evidence-based healthcare practice.

SECTION 9: STUDY LIMITATIONS AND FUTURE DIRECTIONS

Study Limitations and Future Directions

Clinical research studies are invaluable in advancing medical knowledge and informing evidence-based practice. However, it is important to recognize that these studies are not without limitations. Understanding and addressing these limitations is crucial to ensure the validity, reliability, and generalizability of study findings. This section will provide a thorough discussion on the inherent limitations in clinical research studies and offer guidance on strategies to minimize or account for these limitations. Common sources of bias, such as selection bias, information bias, and confounding, will be defined, and strategies to address them during study design and analysis will be explored. Additionally, the importance of acknowledging limitations and discussing their potential impact on study findings and generalizability will be emphasized. Lastly, the section will encourage readers to reflect on the implications of their research findings, identify potential areas for future research, and emphasize the importance of disseminating study results.

Limitations in Clinical Research Studies

Selection Bias: Selection bias occurs when the study sample is not representative of the target population. This can lead to results that are not generalizable or applicable to the broader population. To address selection bias, researchers should strive to use randomization or other appropriate sampling methods to ensure a representative sample. Additionally, careful consideration should be given to eligibility criteria and recruitment strategies to minimize potential biases.

Information Bias: Information bias, also known as measurement bias, occurs when there are errors or inaccuracies in the measurement of study variables. This can result in misclassification or misinterpretation of the data. To minimize information bias, researchers should use standardized and validated measurement tools, provide clear instructions to participants and data collectors, and employ quality control measures throughout the data collection process.

Confounding: Confounding occurs when the relationship between the exposure and outcome is distorted by the influence of a third variable. Confounding variables can introduce bias and affect the interpretation of study findings. To address confounding, researchers should carefully identify potential confounders and consider them during study design and analysis. This can be done through stratification, matching, or statistical adjustment using techniques such as multivariable regression.

Strategies to Address Limitations

Study Design: A well-designed study can help minimize limitations. Researchers should carefully consider the research question, study population, sample size calculation, and appropriate study design (e.g., randomized controlled trials, cohort studies, case-control studies) based on the research objectives. Randomization, blinding, and control groups in experimental designs can help control for bias and confounding.

Standardization: Standardization of protocols, data collection methods, and measurement tools across study sites can minimize variability and enhance the reliability of data. This can be achieved through detailed study protocols, training of study personnel, and adherence to established guidelines and procedures.

Data Collection and Monitoring: Implementing robust data collection methods and conducting regular monitoring and quality checks can help ensure the accuracy and completeness of data. Researchers should invest in comprehensive training for data collectors, implement data validation checks, and monitor data collection processes to minimize errors and inconsistencies.

Statistical Analysis: Appropriate statistical analysis techniques should be employed to account for potential biases and confounding variables. Advanced statistical methods, such as propensity score matching or sensitivity analysis, can help address limitations associated with confounding and selection bias. Consulting with biostatisticians or statisticians experienced in clinical research can enhance the rigor and validity of the statistical analysis.

Acknowledging Limitations and Implications

It is crucial for researchers to acknowledge and discuss the limitations of their studies. This demonstrates transparency and allows readers to critically evaluate the findings. Researchers should clearly state the limitations in their research articles, including any potential sources of bias or confounding. Additionally, discussing the implications of the limitations on the study findings and their generalizability helps to contextualize the results and interpret their real-world significance.

Identifying Areas for Future Research

Recognizing the limitations in a study can open avenues for future research. Researchers should reflect on the gaps or unanswered questions that arise from their findings. These can serve as starting points for future studies and investigations, contributing to the advancement of scientific knowledge in the field.

Dissemination of Study Results

Lastly, researchers should prioritize the dissemination of their study results to ensure the wider impact of their findings. Publishing research articles in peer-reviewed journals, presenting at conferences, and engaging in knowledge translation activities can facilitate the sharing of knowledge and promote evidence-based practice. Transparent reporting of study methods, results, and limitations enhances the reproducibility and trustworthiness of research findings.

Conclusion

Clinical research studies are not without limitations, but by understanding and addressing these limitations, researchers can enhance the validity, reliability, and generalizability of their findings. Selection bias, information bias, and confounding are common sources of bias in clinical research, and various strategies can be employed to minimize or account for them. Acknowledging limitations, identifying areas for future research, and disseminating study results are essential steps in advancing scientific knowledge and promoting evidence-based healthcare practice. By employing rigorous study designs, robust data collection and analysis techniques, and transparent reporting, researchers can strengthen the quality and impact of their clinical research studies.

5

PARTICIPANT RECRUITMENT AND INFORMED CONSENT IN CLINICAL RESEARCH

CHAPTER V

PARTICIPANT RECRUITMENT AND INFORMED CONSENT IN CLINICAL RESEARCH

This chapter focuses on the crucial role that participants play in clinical research. We will discuss the reasons why individuals choose to participate in clinical studies and the impact their involvement has on the success of research endeavors. We will also emphasize the importance of informed consent, confidentiality, and privacy protection for research participants.

SECTION 1: IMPORTANCE OF PARTICIPANT RECRUITMENT

Importance of Participant Recruitment

Participant recruitment is a critical aspect of clinical research that plays a pivotal role in determining the success and validity of a study. The recruitment process involves identifying, screening, and enrolling individuals who meet specific criteria to participate in research investigations. It is crucial to recognize the significance of participant recruitment as it directly impacts the availability and diversity of participants, influencing the generalizability and validity of research findings. This section aims to emphasize the importance of proactive and thoughtful participant recruitment strategies in achieving representative and sufficient sample sizes while addressing the challenges researchers often encounter in this process.

The Role of Participant Recruitment in Study Success

The success of a clinical research study depends on the recruitment of an appropriate and diverse group of participants. Adequate recruitment ensures that the study has a sufficiently large sample size, which is essential for generating statistically significant results and reliable conclusions. Without an adequate number of participants, the study may lack the statistical power required to detect meaningful effects or associations accurately.

Impact on Generalizability and Validity

Participant recruitment is vital in ensuring the generalizability of research findings. By recruiting a diverse range of participants from various demographic backgrounds, researchers can enhance the external validity of their study. This diversity allows for the examination of potential variations in treatment effects across different populations, making the findings more applicable to real-world scenarios.

Moreover, the validity of research findings relies on recruiting participants who accurately represent the target population of interest. By recruiting a representative sample, researchers increase the likelihood that their findings are applicable to the broader population and minimize the potential for selection bias, which can undermine the internal validity of a study.

Challenges in Participant Recruitment

Despite the importance of participant recruitment, researchers often encounter various challenges throughout the process. These challenges can hinder the timely and successful completion of a study and impact its overall quality. Let's identify and discuss some common challenges with participant recruitment.

- **Limited Awareness:** Many potential participants may be unaware of clinical research opportunities. Insufficient public awareness and understanding of clinical trials can impede recruitment efforts. Researchers must actively engage in educational initiatives to increase public knowledge about the importance and benefits of participating in research studies.

- **Communication Gaps:** Inadequate communication between researchers and participants can lead to disconnection and reduced engagement. Clear and regular communication is essential.

- **Logistical Constraints:** Clinical research studies often require participants to meet specific eligibility criteria and participate in various study procedures, such as clinic visits, laboratory tests, and interviews. These logistical requirements can pose significant barriers to recruitment, especially for individuals with limited resources, time constraints, or transportation difficulties. Researchers must consider and address these logistical constraints to facilitate participation and ensure inclusivity.

- **Participant Reluctance:** Some individuals may be hesitant or unwilling to participate in research studies due to concerns about potential risks, privacy, or mistrust of the research process. Participant reluctance can slow down recruitment efforts and lead to selection bias if certain demographic groups are systematically underrepresented. Researchers should prioritize building trust, providing transparent information, and addressing participant concerns to mitigate reluctance and improve recruitment rates.

Ensuring Representation and Sufficient Sample Sizes

To achieve representative and diverse participant samples, researchers need to employ proactive and thoughtful recruitment strategies. These strategies may include:

- **Collaborating with Community Partners:** Engaging with community organizations, healthcare providers, and advocacy groups can facilitate participant recruitment. These partnerships allow researchers to access diverse populations and leverage existing trust and relationships within the community.

- **Targeted Outreach and Education:** Conducting targeted outreach campaigns, both online and offline, can help raise awareness about clinical research opportunities. Educational initiatives that provide clear and accessible information about the study's purpose, procedures, and potential benefits can address potential participant concerns and increase recruitment rates.

- **Inclusive Eligibility Criteria:** Researchers should critically evaluate their eligibility criteria to ensure they do not unnecessarily exclude certain populations. Broadening eligibility criteria when appropriate can enhance diversity and improve the external validity of the study findings.

- **Participant Incentives:** Providing reasonable incentives or compensation for participants' time and effort can encourage recruitment and retention. These incentives should be ethically and appropriately determined to avoid coercion and maintain the integrity of the research process.

- **Culturally Competent Approaches:** Researchers must adopt culturally sensitive recruitment approaches that consider the unique perspectives, values, and beliefs of different communities. Customizing recruitment materials and strategies to align with cultural norms and preferences can foster trust and enhance participation.

Conclusion

Participant recruitment is a critical component of clinical research, significantly influencing the success, generalizability, and validity of study findings. Researchers must recognize the challenges associated with recruitment and employ proactive and thoughtful strategies to ensure the representation of diverse populations and achieve sufficient sample sizes. By addressing limited awareness, logistical constraints, and participant reluctance, researchers can improve recruitment rates and enhance the overall quality and impact of their research.

SECTION 2: RECRUITMENT STRATEGIES AND APPROACHES IN CLINICAL RESEARCH

Recruitment Strategies and Approaches in Clinical Research

In this section, we will delve into various recruitment strategies and approaches employed in clinical research. We will discuss traditional methods, such as recruitment through healthcare providers, clinics, and advertisements, as well as emerging strategies, including online platforms, social media, and community engagement. Readers will gain insights into the advantages and limitations of different recruitment approaches and understand the importance of tailoring strategies to the target population and research objectives. We will provide practical tips and examples to guide researchers in developing effective participant recruitment plans.

Introduction

Implementing effective participant recruitment strategies is essential for successful clinical research. Researchers employ a variety of approaches, ranging from traditional methods to emerging strategies, to reach and engage potential participants. This section explores different recruitment strategies and approaches, highlighting their advantages, limitations, and the importance of tailoring these strategies to the target population and research objectives. Practical tips and examples will be provided to guide researchers in developing effective participant recruitment plans.

Traditional Recruitment Methods

Healthcare Providers and Clinics: Collaborating with healthcare providers and clinics is a common approach to participant recruitment. These settings allow researchers to access individuals who may already be seeking medical care and are therefore more likely to be interested in research participation. Establishing relationships with healthcare professionals, providing study materials, and offering training on the study protocol can enhance recruitment success.

Advertisements: Placing advertisements in newspapers, magazines, radio, and television can reach a broader audience. Researchers can design advertisements that highlight the study's purpose, eligibility criteria, and potential benefits to attract potential participants. However, this method may not always reach specific target populations, and recruitment costs can be significant.

Direct Mail: Sending recruitment materials, such as letters or brochures, directly to potential participants can be an effective strategy. Researchers can obtain mailing lists from various sources, such as patient registries or healthcare organizations. Personalized and informative materials can increase the likelihood of response and participation.

Emerging Recruitment Strategies

Online Platforms: The widespread use of the internet has created new opportunities for participant recruitment. Researchers can leverage online platforms, such as dedicated study websites, research registries, or online communities, to reach a large and diverse pool of potential participants. These platforms allow for convenient dissemination of study information and online screening, simplifying the recruitment process. Online platforms also enable researchers to collect data remotely, enhancing participant accessibility and reducing logistical constraints.

Social Media: Social media platforms, including Facebook, Twitter, and Instagram, have become powerful tools for participant recruitment. Researchers can create study-specific pages or accounts, target specific demographics through advertising features, and engage with potential participants through posts, discussions, and direct messaging. Social media offers the advantage of rapid information dissemination and the potential for viral reach, but researchers must navigate ethical considerations, such as privacy and data security.

Community Engagement: Engaging with communities directly can foster trust and increase participation rates. Researchers can collaborate with community organizations, leaders, and influencers to raise awareness about the study and address community-specific concerns. Participating in community events, hosting informational sessions, and establishing relationships with key stakeholders can create a sense of partnership and improve recruitment success.

Tailoring Strategies to the Target Population and Research Objectives

To develop effective participant recruitment plans, researchers should consider the characteristics of the target population and align their strategies with the research objectives. Some practical tips include:

Understanding the Target Population: Researchers should conduct thorough research to understand the demographics, preferences, and potential barriers of the target population. This knowledge allows for the development of tailored recruitment messages, materials, and strategies.

Building Relationships and Partnerships: Establishing partnerships with community organizations, healthcare providers, and advocacy groups can provide access to specific populations and enhance recruitment success. Collaborating with key stakeholders can help researchers understand community dynamics and address cultural or logistical barriers.

Clear and Engaging Study Materials: Recruitment materials should be clear, concise, and engaging. Use language appropriate for the target population, highlight the benefits and importance of the research, and provide information about the study procedures and eligibility criteria. Visual aids, such as infographics or videos, can enhance comprehension and engagement.

Utilizing Multiple Channels: To reach a diverse population, researchers should employ multiple recruitment channels simultaneously. Combining traditional methods with online platforms, social media, and community engagement can maximize the visibility and accessibility of the study.

Continuous Monitoring and Adaptation: Recruitment strategies should be monitored and evaluated regularly to assess their effectiveness. Researchers should be prepared to modify their approaches based on feedback, recruitment rates, and emerging trends to optimize participant enrollment.

Conclusion

Implementing effective participant recruitment strategies is crucial for the success of clinical research. Researchers can employ a range of traditional and emerging approaches to reach and engage potential participants. By tailoring strategies to the target population and research objectives, researchers can maximize recruitment success. Understanding the advantages, limitations, and practical tips associated with different recruitment methods allows researchers to develop comprehensive and effective participant recruitment plans, ultimately contributing to the overall quality and impact of their research.

SECTION 3: UNDERSTANDING CONSENT FORMS AND REGULATORY REQUIREMENTS IN CLINICAL RESEARCH

Understanding Consent Forms and Regulatory Requirements in Clinical Research

The informed consent process is a cornerstone of ethical research, ensuring that participants have a comprehensive understanding of the study's purpose, risks, benefits, and procedures before deciding to participate. This section provides an extensive exploration of the informed consent process in clinical research, highlighting its importance, key elements, ethical and legal considerations, and practical guidance for conducting the process effectively.

Purpose of Consent Forms

Consent forms serve as a crucial tool to facilitate the informed consent process. They provide participants with detailed information about the research study, enabling them to make an autonomous and well-informed decision regarding their participation. Let's discuss some purposes that consent forms serve.

Ensuring Informed Consent: The consent form ensures that participants are fully informed about the nature, purpose, procedures, risks, benefits, and alternatives associated with the study. This process empowers participants to make an educated decision about their involvement in the trial.

> The consent form provides comprehensive information about the study, including the research objectives, expected duration, detailed procedures (such as blood tests or imaging), potential risks (such as side effects or discomfort), potential benefits (such as access to new treatments), and available alternatives (such as standard treatment options). It ensures that participants understand what they are agreeing to and that their consent is voluntary.

Documenting Voluntary Participation: The consent form serves as a legal document that records the participant's voluntary agreement to take part in the clinical trial. It provides evidence that the participant has willingly decided to participate without any coercion or undue influence.

> Participants sign the consent form to confirm their willingness to join the study. The form includes statements confirming that participation is voluntary, that they have had the opportunity to ask questions, and that they understand they can withdraw from the study at any time without penalty or loss of benefits to which they are otherwise entitled. This documentation is crucial for ethical and legal compliance.

Communicating Participant Rights: The consent form outlines the rights of the participants, ensuring they are aware of their protections and entitlements while participating in the study. It emphasizes their autonomy and the ethical conduct of the research.

> The form details the participant's rights, such as the right to withdraw from the study at any point, the right to privacy and confidentiality of their personal data, and the right to receive a copy of the consent form. It also informs participants about whom to contact if they have questions or concerns about the study or their rights. This transparency fosters trust and ethical integrity.

Ensuring Compliance with Ethical Standards: Consent forms help ensure that clinical trials adhere to ethical guidelines and regulatory requirements. They are a key component of ethical oversight by Institutional Review Boards (IRBs) or Ethics Committees.

> Ethical guidelines, such as the Declaration of Helsinki and the Belmont Report, mandate the use of informed consent to protect participants. Consent forms provide a structured way to meet these ethical standards. They are reviewed and approved by IRBs to ensure that the information is comprehensive, understandable, and ethically sound. This oversight helps protect participants from harm and exploitation.

Facilitating Transparency and Trust: The use of consent forms promotes transparency between researchers and participants, fostering trust in the research process. It ensures that participants are aware of all aspects of the study and can make informed decisions.

> By providing clear, detailed information about the study, consent forms help build a relationship of trust between researchers and participants. Participants who understand the study are more likely to trust the researchers and feel respected and valued.

Key Components of Consent Forms

To ensure clarity and ensure participant comprehension, consent forms typically include the following components:

- **Study Information:** Consent forms provide a concise summary of the research study, including its purpose, objectives, study procedures, and duration. This section helps participants understand the context and scope of the study that they may be signing up for.

- **Risks and Benefits:** Researchers outline the potential risks and benefits associated with participation. This includes both physical and psychological risks, as well as any potential benefits to the participant or society as a whole. Clear and accurate information allows individuals to assess the potential risks and benefits of their involvement.

- **Confidentiality and Privacy:** Consent forms explain how participants' confidentiality and privacy will be protected. Researchers should outline the measures taken to ensure the security and anonymity of collected data, along with any limitations or exceptions to confidentiality, such as mandatory reporting requirements.

- **Voluntary Participation and Withdrawal:** Consent forms explicitly state that participation is voluntary and that individuals have the right to withdraw from the study at any time without penalty or loss of benefits. Procedures for withdrawing and the implications of withdrawal should be clearly described.

- **Contact Information:** Consent forms provide contact information for the research team, including the principal investigator or study coordinator. Participants should be aware of whom to contact for any questions, concerns, or clarifications regarding the study.

- **Signature and Date:** Consent forms require the participant's signature and the date of consent to indicate their willingness to participate after understanding the information provided. For individuals who may have limited literacy or language proficiency, alternative methods of documenting consent should be considered.

Regulatory Requirements for Consent Forms

Consent forms in clinical research must adhere to ethical principles and regulatory requirements, such as the following:

- **Ethical Principles:** Consent forms should uphold the core ethical principles of respect for autonomy, beneficence, and justice. Participants must be given sufficient information to make informed decisions, and their rights, welfare, and privacy should be protected.

- **Institutional Review Board (IRB) Approval:** Prior to participant recruitment, researchers must submit their consent forms to an independent IRB or ethics committee for review. The IRB ensures that the study complies with ethical standards and regulatory guidelines, and that the consent form is comprehensive and understandable.

- **Informed Consent Process:** Consent forms are accompanied by an informed consent process that involves a verbal explanation of the study and an opportunity for participants to ask questions and clarify any concerns. The consent form should complement this process and not serve as a substitute for active communication between researchers and participants.

- **Regulatory Guidelines:** Different countries have specific regulatory guidelines regarding informed consent in clinical research. Researchers must comply with applicable regulations, such as the Declaration of Helsinki, the Common Rule (U.S.), and the European Union's General Data Protection Regulation (GDPR). These guidelines stipulate the information that must be included in consent forms and the processes for obtaining and documenting consent.

Ensuring Participant Autonomy and Protection

Obtaining informed consent is essential to respect participant autonomy and protect their rights and welfare. To ensure participant autonomy:

- **Clarity and Comprehensibility:** Consent forms should be written in clear, concise, and understandable language. Technical jargon should be avoided, and complex concepts should be explained in lay terms. Consideration should be given to participants with different literacy levels and cultural backgrounds.

- **Ongoing Communication:** Consent forms should be accompanied by ongoing communication between researchers and participants throughout the study. Participants should have access to a contact person who can address any questions or concerns that may arise during their participation.

- **Capacity Assessment:** Researchers should assess participants' capacity to provide informed consent. Individuals with cognitive impairments, language barriers, or diminished decision-making abilities may require additional support, such as the involvement of a legally authorized representative or a qualified interpreter.

- **Consent for Minors and Vulnerable Populations:** Special considerations apply when obtaining consent from minors and vulnerable populations, such as individuals with disabilities or those who are socioeconomically disadvantaged. Researchers must comply with specific regulations and obtain additional consent or assent from appropriate parties.

Key Elements of the Informed Consent Process

The informed consent process in clinical research encompasses several key elements to ensure that participants are adequately informed and can make autonomous decisions. These elements include:

- **Voluntariness:** Participants must provide consent voluntarily, without coercion or undue influence. They should feel free to decline participation or withdraw from the study at any time without facing negative consequences.

- **Comprehension:** Participants should have a clear understanding of the study's purpose, procedures, potential risks and benefits, and any alternative treatment options. Researchers should use lay language, avoid technical jargon, and provide opportunities for participants to ask questions and seek clarification.

- **Disclosure of Information:** Researchers must provide comprehensive and accurate information to participants. This includes disclosing any known risks, potential benefits, anticipated discomforts, confidentiality measures, and any potential conflicts of interest.

- **Capacity to Provide Consent:** Participants must have the capacity to provide informed consent, meaning they possess the cognitive ability to understand and evaluate the information provided. Researchers should assess participants' decision-making capacity, especially when working with vulnerable populations or individuals with cognitive impairments.

Ethical and Legal Considerations

The informed consent process in clinical research is subject to ethical principles and legal requirements. Researchers must consider the following:

- **Documentation:** Informed consent should be documented through a consent form signed by the participant or their legally authorized representative. The consent form should include key study information, risks, benefits, confidentiality measures, and the voluntary nature of participation. It serves as proof that informed consent was obtained.

- **Ongoing Consent:** Informed consent is not a one-time event but an ongoing process throughout the study. Researchers should ensure that participants are informed of any significant changes to the study protocol or potential risks and benefits that may emerge during the course of the research.

- **Vulnerable Populations:** Special considerations apply when obtaining informed consent from vulnerable populations, such as children, individuals with cognitive impairments, or those who are socioeconomically disadvantaged. Additional safeguards and procedures may be required to protect the rights and welfare of these individuals.

- **Research Ethics Review:** The informed consent process, including consent forms, must undergo review by an independent Institutional Review Board (IRB) or ethics committee. These bodies evaluate the ethical soundness of the research and ensure that informed consent procedures meet regulatory guidelines.

Conducting the Informed Consent Process Effectively

To conduct the informed consent process effectively, researchers should consider the following guidelines:

- **Clear Communication:** Researchers should communicate with participants using clear and understandable language, avoiding technical jargon. They should provide information in a manner that matches participants' literacy levels and cultural backgrounds.

- **Individualized Approach:** Researchers should tailor the consent process to each participant, considering their specific needs, concerns, and comprehension level. They should allow sufficient time for participants to absorb information and ask questions.

- **Addressing Questions and Concerns:** Researchers should encourage participants to ask questions and address any concerns they may have. They should provide accurate and honest responses, acknowledging uncertainties when necessary.

- **Supplementary Materials:** Researchers can supplement the consent form with additional materials, such as information sheets, visual aids, or multimedia presentations, to enhance participant understanding.

- **Multidisciplinary Collaboration:** Collaboration between researchers and professionals from diverse backgrounds, such as ethics consultants, community representatives, or interpreters, can enhance the effectiveness of the consent process, especially when working with culturally diverse populations or individuals with language barriers.

- **Confidentiality and Privacy:** Researchers should emphasize the measures taken to protect participants' confidentiality and privacy throughout the study. They should clarify any limitations or exceptions to confidentiality, such as mandatory reporting requirements, and address any concerns participants may have regarding data security.

- **Participant Decision Reassessment:** Researchers should periodically reassess participants' understanding and willingness to continue participating in the study. This allows participants to reevaluate their decisions and withdraw if they wish to do so.

Conclusion

Consent forms are vital components of the informed consent process in clinical research. They serve to provide participants with detailed information, ensure their autonomy, and protect their rights and welfare. By adhering to regulatory requirements, maintaining clear and comprehensive communication, and prioritizing participant understanding, researchers can foster trust and ethical conduct in their studies. A well-designed and properly executed consent form contributes to the integrity and credibility of clinical research while upholding the principles of autonomy and participant protection.

SECTION 4: SPECIAL CONSIDERATIONS IN PARTICIPANT RECRUITMENT

Special Considerations in Participant Recruitment

Participant recruitment for specific populations or study types requires tailored strategies and considerations to ensure representation, diversity, and successful enrollment. This section explores the challenges and recruitment strategies for diverse populations, including children, older adults, individuals from minority backgrounds, and individuals with limited health literacy. It also addresses recruitment considerations for specific study types, such as genetic research, clinical trials, and studies involving sensitive topics. The emphasis is on cultural sensitivity, trust-building, and rapport establishment to overcome barriers and enhance recruitment success.

Recruitment Challenges and Strategies for Diverse Populations

Children: Recruiting children for research involves additional ethical and legal considerations. Challenges include obtaining parental or guardian consent, assessing the child's capacity to provide assent, and addressing potential fears or discomfort. Strategies include involving pediatricians or child specialists, using child-friendly language and materials, and engaging parents or guardians as active partners in the recruitment process.

Older Adults: Recruiting older adults may be challenging due to health issues, cognitive decline, or limited mobility. Strategies include collaborating with geriatric specialists or healthcare providers, providing clear and accessible information, accommodating physical needs, and conducting home visits if necessary. Building trust and addressing concerns about privacy and confidentiality are essential.

Individuals from Minority Backgrounds: Recruiting individuals from minority backgrounds requires cultural sensitivity and efforts to establish trust. Strategies involve partnering with community organizations, engaging community leaders or influencers, using culturally appropriate recruitment materials, and involving bilingual or culturally competent staff. Building relationships and addressing concerns about representation and inclusivity are crucial.

Individuals with Limited Health Literacy: Recruiting individuals with limited health literacy necessitates clear and simplified communication. Strategies include using plain language, visual aids, and multimedia resources, ensuring adequate time for comprehension and questions, and involving healthcare professionals or patient advocates to assist with the recruitment process. Ensuring a supportive and non-judgmental environment is vital.

Recruitment Considerations for Specific Study Types

Genetic Research: Recruiting participants for genetic research involves specific considerations due to the complex nature of genetics. Strategies include collaborating with genetic counselors or specialists, providing genetic education and counseling services, addressing privacy and confidentiality concerns, and highlighting the potential benefits of genetic research, such as personalized medicine or familial health insights.

Clinical Trials: Recruiting participants for clinical trials requires addressing concerns about experimental treatments, potential risks, and randomization. Strategies include engaging with healthcare providers and patient advocacy groups, disseminating information through multiple channels, providing clear explanations of the trial's purpose and procedures, and emphasizing the potential contribution to scientific advancements or improved treatments.

Studies Involving Sensitive Topics: Recruiting participants for studies involving sensitive topics, such as mental health, substance abuse, or stigmatized conditions, requires creating a safe and non-judgmental environment. Strategies involve partnering with mental health professionals or support organizations, ensuring strict confidentiality measures, using anonymous data collection methods if applicable, and emphasizing the potential societal impact of the study.

Practical Recommendations for Overcoming Barriers and Enhancing Recruitment Success

Cultural Sensitivity: Tailor recruitment materials and approaches to the cultural context and preferences of the target population. Consider language, symbols, traditions, and cultural nuances to foster trust and engagement.

Building Trust and Rapport: Engage with community leaders, influencers, or organizations to establish trust and credibility. Conduct outreach events, establish long-term relationships, and emphasize the value of participants' contributions.

Accessibility and Convenience: Ensure that the recruitment process is convenient and accessible for participants. Offer flexible scheduling, convenient locations, transportation assistance if needed, and accommodate any special needs or considerations.

Clear Communication: Use plain language and visual aids to enhance comprehension. Provide ample opportunities for participants to ask questions, and ensure that responses are clear, accurate, and culturally appropriate.

Incentives and Benefits: Consider offering appropriate incentives or benefits to participants, such as reimbursement for travel expenses, compensation for time, or access to study-related resources or interventions. Ensure transparency and fairness in incentive distribution.

Empowerment and Participant Engagement: Involve participants as active partners in the research process. Solicit their input, involve them in study design or dissemination activities, and provide opportunities for them to share their experiences and perspectives.

Conclusion

Recruiting participants from specific populations or for specific study types requires tailored strategies and considerations to overcome challenges and enhance recruitment success. By adopting culturally sensitive approaches, building trust, and addressing specific barriers, researchers can ensure the representation, diversity, and active engagement of these populations in research studies. The result is a more inclusive and valid research process that respects the unique needs and perspectives of diverse populations.

SECTION 5: PARTICIPANT RETENTION AND ENGAGEMENT

Participant Retention and Engagement

Participant retention and engagement are critical aspects of conducting successful clinical research. Maintaining participants' involvement throughout the study duration is essential to minimize attrition rates, maximize data completeness, and ensure the validity and reliability of research findings. Understanding the factors that influence retention and implementing effective strategies can enhance data quality and ensure robust research outcomes. This section will delve into various aspects that contribute to the significance of participant retention and engagement in clinical research.

The Significance of Participant Retention and Engagement

Data Integrity and Study Validity: Retaining participants throughout the study duration is essential for maintaining data integrity and ensuring the validity of the research findings. High attrition rates can lead to incomplete data sets, which may compromise the statistical power and generalizability of the study results. Consistent follow-up with participants helps in minimizing the risk of data loss and ensures that the study outcomes accurately reflect the intervention's effects.

Cost Efficiency: Participant retention is also a cost-efficient strategy. Recruiting new participants to replace those who drop out can be expensive and time-consuming. By focusing on retention, researchers can avoid the additional costs associated with re-recruitment and re-consent processes, thereby optimizing resource utilization.

Statistical Power: High participant retention rates contribute to increased statistical power, allowing researchers to detect smaller effect sizes and draw more robust conclusions. It enhances the study's ability to detect significant differences between groups or interventions.

Representativeness: Adequate participant retention promotes the representativeness of the study sample. High attrition rates can introduce bias if participants who drop out differ systematically from those who remain, potentially impacting the generalizability of the findings.

Ethical Considerations: Researchers have an ethical obligation to optimize participant retention. Participants invest their time and effort in research studies, and respecting their commitment involves minimizing attrition and maximizing the value of their contributions.

Challenges in Participant Retention and Engagement

Participant Burden: Study-related procedures, such as frequent clinic visits, invasive tests, or lengthy questionnaires, can be burdensome and lead to participant fatigue or disengagement. Minimizing participant burden and optimizing study procedures can help alleviate this challenge.

Loss of Interest or Motivation: Participants may lose interest or motivation over time, particularly in long-term studies. Researchers should strive to maintain participants' enthusiasm by providing regular study updates, emphasizing the importance of their contributions, and demonstrating the impact of their involvement.

Time Constraints and Conflicting Commitments: Participants may face competing priorities and time constraints, leading to difficulties in maintaining their engagement with the study. Researchers should be flexible in scheduling study visits, provide reminders, and accommodate participants' schedules as much as possible.

Communication Gaps: Inadequate communication between researchers and participants can contribute to a sense of disconnection and reduced engagement. Regular and clear communication is essential to keep participants informed about study progress, upcoming activities, and any relevant updates.

Strategies for Enhancing Participant Engagement and Retention

Clear and Regular Communication: Effective communication is paramount in keeping participants engaged. Researchers should maintain regular contact with participants, providing updates on the study's progress and addressing any concerns promptly. Clear, transparent communication fosters trust and helps participants feel valued and informed about their contributions to the research. Utilize various communication channels, such as email, phone calls, newsletters, or secure online portals, to ensure consistent contact.

Study Updates and Progress Reports: Share periodic updates on the study's progress, milestones achieved, and key findings (if appropriate). Highlight the significance of participants' contributions and emphasize the impact of their involvement on advancing knowledge and improving healthcare.

Incentives and Acknowledgment: Offering incentives can be a powerful motivator for participants to remain in the study. Incentives can range from financial compensation to non-monetary rewards such as gift cards, personalized feedback, or health check-ups. Acknowledging participants' efforts and contributions can enhance their commitment to the study. Ensure transparency and fairness in the distribution of incentives.

Building a Supportive Research Environment: Create a supportive and inclusive research environment where participants feel valued, respected, and comfortable. Foster a sense of community among participants through support groups, social events, or online forums, facilitating peer-to-peer interactions and support.

Participant-Centered Approach: Adopting a participant-centered approach involves considering the needs and preferences of the participants. Flexibility in scheduling appointments, providing transportation assistance, and creating a supportive research environment can significantly improve retention rates. Personalized interactions and a respectful attitude towards participants' time and contributions also play a vital role in engagement.

Support Systems: Providing access to support systems, such as counseling services or peer support groups, can help participants cope with the demands of the study. This is particularly important in long-term studies or those involving vulnerable populations. Emotional and psychological support can mitigate feelings of burden and enhance participants' willingness to continue.

Tracking and Reminders: Implement systems to track participant progress, send reminders for upcoming study activities or visits, and ensure participants stay engaged throughout the study duration. Automated reminders via email, text messages, or mobile applications can be effective in reducing missed appointments.

Addressing Participant Concerns: Actively listen to participants' concerns and address them promptly and transparently. Provide opportunities for participants to voice their questions or grievances and ensure that their feedback is taken into account.

Monitoring and Adapting Retention Strategies

Regular Assessment: Regular assessment of retention strategies is crucial for identifying potential issues early and making necessary adjustments. Researchers should monitor retention rates, gather feedback from participants, and analyze the effectiveness of current engagement methods. This proactive approach helps in refining strategies to better meet participants' needs.

Flexibility and Adaptation: Flexibility in study protocols and willingness to adapt to participants' circumstances can improve retention. This might involve modifying follow-up schedules, offering remote participation options, or simplifying study procedures. Being responsive to participants' feedback and demonstrating adaptability can foster a positive research experience and encourage continued participation.

Conclusion

Participant retention and engagement play a vital role in the success of clinical research studies. By recognizing the significance of participant involvement, understanding the challenges faced, and implementing strategies to enhance retention and engagement, researchers can optimize data completeness, improve statistical power, and uphold ethical obligations. Building a supportive research environment, maintaining regular communication, and offering incentives and acknowledgments contribute to fostering long-term participant engagement and ensuring the validity and generalizability of research findings.

SECTION 6: ETHICAL CONSIDERATIONS IN PARTICIPANT RECRUITMENT

Ethical Considerations in Participant Recruitment

Ethical considerations play a crucial role in participant recruitment for clinical research. Ensuring fairness, respect for autonomy, privacy, and the protection of vulnerable populations are fundamental principles that guide the recruitment process. This section examines the importance of ethical considerations in participant recruitment, explores potential challenges and pitfalls, and highlights guidelines and regulations governing participant recruitment. By maintaining ethical vigilance, researchers can uphold the rights and well-being of research participants.

Fairness and Equal Opportunity

Avoiding Discrimination: Researchers should ensure that recruitment processes are free from discrimination based on characteristics such as race, ethnicity, gender, age, socioeconomic status, or disability. All individuals meeting the study criteria should have an equal opportunity to participate.

Inclusive Recruitment: Efforts should be made to include diverse populations, ensuring representation, and promoting equitable access to research opportunities. Special attention should be given to historically underrepresented groups to address healthcare disparities and ensure the generalizability of research findings.

Informed Consent

Informed Consent Process: The informed consent process should adhere to ethical guidelines and regulations. It should involve clear communication, comprehension assessment, provision of adequate time for decision-making, and addressing any questions or concerns raised by participants. Documentation of informed consent is essential.

Comprehensive Information: Participants must be provided with clear, comprehensive information about the study's purpose, procedures, risks, and benefits. This ensures that they can make an informed decision about their participation.

Voluntary Participation: Participation should be entirely voluntary, with no coercion or undue influence. Participants must understand that they can withdraw from the study at any time without penalty. Researchers should ensure that participants have sufficient information about the study, including its purpose, risks, benefits, and alternatives, to make an autonomous decision.

Respect for Autonomy and Decision-Making

Empowering Participants: Researchers must respect participants' autonomy by ensuring they have all necessary information to make informed decisions about their involvement.

Capacity to Provide Consent: Researchers must assess the decision-making capacity of potential participants, particularly in cases involving vulnerable populations or individuals with impaired decision-making abilities. Additional safeguards may be necessary to protect those who are unable to provide informed consent independently. This ensures that only those who fully understand the implications of participation are enrolled.

Privacy and Confidentiality

Privacy Protection: Researchers must ensure that participants' privacy is protected throughout the recruitment process. Personal identifying information should be handled securely, and data should be anonymized or de-identified whenever possible to maintain confidentiality.

Confidentiality Measures: Researchers should clearly communicate the measures taken to protect participant confidentiality, including data storage, access restrictions, and the use of coded identifiers. Participants should be informed about any exceptions to confidentiality, such as mandatory reporting requirements.

Protection of Vulnerable Populations

Special Consideration: Researchers must give special attention to vulnerable populations, including children, pregnant women, individuals with cognitive impairments, prisoners, and economically disadvantaged individuals. Additional safeguards may be necessary to protect their rights, welfare, and privacy.

Independent Advocacy: In some cases, involving independent advocates or guardians may be necessary to ensure the well-being and rights of vulnerable participants. These advocates should act in the best interests of the participants and serve as a safeguard against potential exploitation.

Conflicts of Interest

Transparency and Disclosure: Researchers should disclose any potential conflicts of interest that may influence the recruitment process. This includes financial interests, affiliations, or relationships that may compromise the objectivity, fairness, or integrity of the study.

Minimizing Coercion and Undue Influence: Researchers must be cautious not to exert undue influence or coercion during the recruitment process. Participants should be given sufficient time and information to make an autonomous decision, free from external pressures.

Conclusion

Ethical considerations in participant recruitment are vital to protect the rights, autonomy, and well-being of research participants. Adhering to principles of fairness, respect for autonomy, privacy, and protection of vulnerable populations is essential. Researchers should be guided by ethical guidelines and regulations, ensuring an inclusive and transparent recruitment process. By maintaining ethical vigilance, researchers can uphold the integrity of the study and contribute to the advancement of scientific knowledge while respecting the rights and well-being of research participants.

SECTION 7: PARTICIPANT FEEDBACK AND DISSEMINATION OF RESULTS

Participant Feedback and Dissemination of Results

Participant feedback and the dissemination of study results are crucial components of ethical and responsible clinical research. Researchers have an ethical obligation to share study findings with participants, ensuring transparency and fostering a sense of partnership and respect. This section emphasizes the importance of providing participant feedback and discusses the potential benefits of participant engagement in improving research practices. Various approaches to disseminating study results, such as individualized reports, community presentations, and lay summaries, will be explored. Ethical and practical considerations in disseminating study results to participants and the broader community will be addressed.

Ethical Obligation to Provide Participant Feedback

Respect for Participant Autonomy: Sharing study findings with participants respects their autonomy and acknowledges their contribution to the research process. Participants have a right to be informed about the outcome of their involvement and the broader impact of the study.

Transparency and Trust: Providing participant feedback enhances transparency and fosters trust between researchers and participants. It demonstrates a commitment to open communication and recognizes participants as valued partners in the research endeavor.

Empowerment and Engagement: Participant feedback empowers individuals by providing them with knowledge and information about the research they participated in. It encourages their ongoing engagement, facilitates informed decision-making, and may stimulate interest in future research opportunities.

Benefits of Participant Feedback

Participant Satisfaction and Retention: Providing feedback to participants contributes to their overall satisfaction with the research experience. It demonstrates that their involvement was meaningful and valued, which can increase participant retention rates and improve future recruitment efforts.

Quality Improvement: Participant feedback can provide valuable insights into the research process, identifying areas for improvement. Participants' perspectives and suggestions can guide researchers in refining study protocols, enhancing data collection methods, and improving the overall participant experience.

Scientific Advancement: Participant feedback can lead to new research questions and avenues for exploration. Participants may offer unique perspectives, personal experiences, or insights that could shape future studies and contribute to scientific advancements.

Approaches to Providing Participant Feedback

Individualized Reports: Researchers can provide individualized reports to participants, summarizing the study findings in a personalized manner. These reports can be tailored to the participants' level of understanding and may include explanations of key results and their implications.

Community Presentations: Organizing community presentations or town hall meetings allows researchers to disseminate study results to a broader audience, including participants, their families, and the local community. This approach promotes engagement, facilitates discussion, and encourages knowledge exchange.

Lay Summaries and Plain Language Materials: Researchers can develop lay summaries or plain language materials that distill complex study findings into accessible and easily understandable formats. These materials are particularly useful for participants with diverse backgrounds and varying levels of scientific literacy.

Digital and Online Platforms: Utilizing digital platforms, such as study websites, social media, or online forums, can facilitate the dissemination of study results to participants. These platforms offer interactive features that encourage feedback, questions, and discussions.

Ethical and Practical Considerations

Timeliness: Researchers should strive to provide participant feedback in a timely manner to ensure that the information remains relevant and meaningful. Delays in sharing results may diminish the impact and participants' interest in the findings.

Clear and Understandable Language: When disseminating study results, researchers should use clear and jargon-free language to ensure participants can comprehend and interpret the information accurately. The use of visual aids, graphics, and infographics can enhance understanding.

Privacy and Confidentiality: Researchers must consider privacy and confidentiality when disseminating study results. Ensure that individual participants cannot be identified through the provided feedback and that sensitive or personally identifiable information is appropriately safeguarded.

Two-Way Communication: Encourage participants to provide feedback, ask questions, and engage in discussions about the study findings. Establishing a two-way communication channel allows for ongoing dialogue and strengthens the researcher-participant relationship.

Cultural Sensitivity and Accessibility: Take into account cultural diversity and accessibility needs when disseminating study results. Use culturally appropriate language and consider providing translated materials or using alternative formats to reach diverse populations.

Conclusion

Participant feedback and the dissemination of study results are integral components of ethical and responsible clinical research. Researchers have an ethical obligation to share study findings with participants, fostering transparency, trust, and ongoing engagement. Participant feedback can improve research practices, enhance participant satisfaction, and contribute to scientific advancement. By utilizing various approaches and considering ethical and practical considerations, researchers can effectively disseminate study results to participants, empowering them with knowledge and promoting meaningful participation in the research process.

6

DATA COLLECTION AND MANAGEMENT IN CLINICAL RESEARCH

CHAPTER VI

DATA COLLECTION AND MANAGEMENT IN CLINICAL RESEARCH

Chapter 6 delves into the critical aspects of data collection and management in clinical research. This chapter aims to provide researchers with a comprehensive understanding of the methods and best practices for collecting high-quality data, ensuring its accuracy, and managing it efficiently throughout the research process. By exploring various data sources, designing robust data collection instruments, and implementing effective data management strategies, researchers can enhance the reliability and validity of their study findings. This chapter also emphasizes the importance of data quality assurance, privacy, confidentiality, and security to uphold the integrity of clinical research. Through this in-depth exploration, readers will gain the knowledge and tools necessary to execute rigorous and ethically sound clinical research.

SECTION 1: OVERVIEW OF DATA COLLECTION IN CLINICAL RESEARCH

Overview of Data Collection in Clinical Research

Data collection is a critical component of clinical research, as it forms the foundation for generating valid and reliable study findings. This section provides an overview of data collection in clinical research, highlighting the importance of high-quality data and the various sources commonly utilized. By understanding the advantages, limitations, and considerations associated with different data sources, researchers can make informed decisions and select appropriate methods for data collection.

Importance of High-Quality Data

Validity and Reliability: Collecting high-quality data is essential to ensure the validity and reliability of study findings. Valid data accurately measure the intended concepts or variables, while reliable data yield consistent results upon repeated measurements.

Evidence-Based Decision Making: Reliable data serve as the basis for evidence-based decision making in clinical practice and policy development. Well-collected and analyzed data contribute to the generation of robust evidence that informs healthcare interventions and improves patient outcomes.

Generalizability: High-quality data enhance the generalizability of research findings, allowing for valid inferences about larger populations. Data collection methods that minimize bias and capture diverse perspectives contribute to the external validity of the study.

Sources of Data in Clinical Research

Medical Records: Medical records provide valuable information on patients' medical history, diagnoses, treatments, and outcomes. Electronic health records (EHRs) have facilitated the accessibility and integration of medical data for research purposes.

Medical Records Advantages:

- **Accessibility:** Medical records are readily available sources of data since they are routinely generated and stored by healthcare providers, making them easily accessible for research purposes.
- **Comprehensive Patient Information:** Medical records contain a wealth of information on patients' medical history, diagnoses, treatments, and outcomes, providing a comprehensive overview of their healthcare journey.
- **Longitudinal Data Analysis:** Medical records often span a significant period, allowing researchers to analyze data longitudinally and examine changes in patient health status over time.

Medical Records Disadvantages:

- **Data Quality Issues:** Medical records may suffer from inconsistencies, errors, or missing data due to variations in documentation practices, different healthcare providers' styles, or human errors during data entry.
- **Missing Data:** Some information may be incomplete or unavailable in medical records, leading to missing data that can affect the overall analysis and interpretation.
- **Variability in Documentation Practices:** Documentation practices may vary across healthcare providers or institutions, leading to differences in the level of detail and quality of information recorded in medical records.

Questionnaires: Questionnaires are self-report tools used to collect information on participants' demographics, symptoms, behaviors, attitudes, and quality of life. They can be administered in person, by mail, or through online platforms.

Questionnaire Advantages:

- **Ease of Administration:** Questionnaires are relatively easy to administer, allowing for efficient data collection from a large number of participants.
- **Standardized Data Collection:** Standardized questionnaires ensure consistent data collection across participants, enabling comparability and reducing potential biases in responses.
- **Large Sample Sizes:** Questionnaires can be distributed widely, reaching a large and diverse population, thereby facilitating the collection of data from a sizable sample for analysis.

Questionnaire Disadvantages:

- **Response Bias:** Participants may provide biased responses due to social desirability, recall bias, or their interpretation of the questions, potentially impacting the validity and accuracy of the data.
- **Recall Bias:** Participants may have difficulty accurately recalling past events, experiences, or behaviors, introducing a potential bias in the responses provided.
- **Reliance on Self-Report:** Questionnaires rely on participants' self-report, which can be influenced by their subjective interpretations, memory, or perception of the questions, leading to potential inaccuracies.

Interviews: Interviews involve direct communication between researchers and participants, allowing for in-depth exploration of topics. Structured, semi-structured, or open-ended interviews can provide rich qualitative data and insights.

<u>Interview Advantages:</u>

- **In-Depth Data Collection:** Interviews allow for in-depth exploration of topics, providing rich qualitative data and insights that cannot be easily captured through other data collection methods.
- **Flexibility:** Researchers can adapt the interview process, follow-up with probing questions, and explore emerging themes, allowing for flexibility and the capture of nuanced information.
- **Capturing Participant Perspectives:** Interviews enable researchers to directly capture participants' perspectives, experiences, and emotions, providing a deeper understanding of their thoughts and motivations.

<u>Interview Disadvantages:</u>

- **Potential Interviewer Bias:** Interviewer bias may unintentionally influence participants' responses, either through leading questions or unintentional cues, compromising the objectivity of the data collected.
- **Resource-Intensive Nature:** Conducting interviews requires time, skilled interviewers, and resources for transcription and data analysis, making it a resource-intensive data collection method.
- **Influence of Social Desirability:** Participants may provide responses influenced by social desirability, seeking to present themselves in a favorable light or conforming to societal expectations, potentially impacting the validity of the data collected.

Laboratory Tests: Laboratory tests involve the analysis of biological samples, such as blood, urine, or tissue, to measure biomarkers, genetic information, or physiological parameters. These tests provide objective and quantitative data.

Laboratory Test Advantages:

- **Objective Measurements:** Laboratory tests provide objective and quantifiable measurements of biomarkers, genetic information, or physiological parameters, reducing the potential for bias in data collection.
- **Precise Quantification:** Laboratory tests often offer precise and accurate measurements, allowing for precise quantification of variables of interest.
- **Assessment of Biological Markers:** Laboratory tests enable the assessment of specific biological markers or indicators that can provide insights into disease processes, treatment responses, or physiological changes.

Laboratory Test Disadvantages:

- **Invasiveness:** Some laboratory tests require invasive procedures, such as blood draws or tissue biopsies, which may be uncomfortable or pose minimal risks to participants.
- **Cost:** Laboratory tests can be costly, especially if specialized equipment, reagents, or trained personnel are required, potentially increasing the overall research budget.
- **Variations in Sample Collection and Analysis:** Differences in sample collection techniques, storage, transportation, and laboratory analysis procedures may introduce variability in results and affect the comparability of data across studies or sites.

Imaging Studies: Imaging techniques, including X-rays, computed tomography (CT), magnetic resonance imaging (MRI), and positron emission tomography (PET), generate visual representations of internal body structures for diagnostic or research purposes.

<u>Imaging Study Advantages:</u>

- **Visualizing Anatomical Structures:** Imaging techniques, such as X-rays, CT scans, MRI, and PET scans, provide visual representations of internal body structures, aiding in the diagnosis, characterization, and monitoring of various conditions.
- **Identification of Pathological Conditions:** Imaging studies can identify pathological conditions, such as tumors, fractures, or abnormalities, contributing to clinical decision-making and research investigations.
- **Non-Invasive Data Collection:** Most imaging studies are non-invasive, avoiding the need for invasive procedures and reducing discomfort for participants.

<u>Imaging Study Disadvantages:</u>

- **High Cost:** Imaging studies can be expensive due to the need for specialized equipment, trained technicians, and data analysis expertise, potentially limiting their feasibility in some research settings.
- **Potential Risks:** Some imaging modalities, such as CT scans or PET scans using radioactive tracers, involve exposure to radiation, which carries potential risks, particularly with repeated exposures.
- **Need for Specialized Equipment and Expertise:** Performing imaging studies requires access to and proficiency in using specialized equipment, as well as the expertise to interpret the obtained images accurately.

Wearable Devices: Wearable devices, such as activity trackers, heart rate monitors, or smartwatches, collect real-time data on physical activity, sleep patterns, vital signs, and other health-related parameters.

Wearable Device Advantages:

- **Continuous and Real-Time Data Collection:** Wearable devices allow for continuous and real-time data collection, providing a more comprehensive picture of participants' behaviors, physiological parameters, or activity patterns.
- **Remote Monitoring:** Wearable devices enable remote monitoring of participants, reducing the need for in-person visits and offering convenience and flexibility.
- **Large-Scale Data Capture:** With the growing popularity of wearable devices, researchers can potentially collect data from large sample sizes, facilitating population-level analysis and insights.

Wearable Device Disadvantages:

- **Accuracy and Reliability Concerns:** The accuracy and reliability of data collected by wearable devices may vary depending on the specific device, calibration, and participant compliance with wearing and using the device as instructed.
- **Limited Validation of Certain Devices:** Some wearable devices may lack rigorous validation or standardization, making it important for researchers to critically evaluate their reliability and validity for the specific research objectives.
- **Potential Participant Burden:** Wearing and maintaining wearable devices may impose a burden on participants, affecting compliance, adherence, and overall participation rates.

Sources of Clinical Data

Please note that these advantages and disadvantages of sources of clinical data are not exhaustive, and additional considerations may apply depending on the specific research context and study design.

Considerations for Selecting Data Collection Methods

- **Research Objectives:** The choice of data collection methods should align with the research objectives and the specific research question being investigated.
- **Feasibility:** Researchers should consider the practicality and resources required for data collection, including personnel, equipment, time, and participant burden.
- **Population Characteristics:** The characteristics of the target population, such as age, cultural background, or health literacy levels, should be taken into account when selecting data collection methods to ensure their appropriateness and acceptability.
- **Ethical Considerations:** Researchers must adhere to ethical principles when collecting data, ensuring informed consent, privacy protection, and the minimization of participant burden or harm.

Conclusion

Data collection is a crucial aspect of clinical research, as it lays the groundwork for generating valid and reliable study findings. By understanding the advantages and limitations of different data sources, researchers can select appropriate methods for data collection. Considerations such as research objectives, feasibility, population characteristics, and ethical considerations should guide the selection process. Collecting high-quality data contributes to evidence-based decision making, enhances research validity, and improves patient outcomes.

SECTION 2: DESIGNING DATA COLLECTION INSTRUMENTS

Designing Data Collection Instruments

Designing data collection instruments, such as questionnaires and interview guides, is a crucial step in ensuring the quality and reliability of data in clinical research. In this section, we will discuss key principles of instrument design, including clarity, relevance, and validity. We will explore various aspects of instrument development, such as selecting appropriate measurement scales, conducting pilot testing, and ensuring cultural and linguistic appropriateness. By following these principles and incorporating practical tips, researchers can design effective data collection instruments that yield accurate and meaningful data, enhancing the overall quality of their research.

Clarity

- **Use Clear and Concise Language:** Data collection instruments should use language that is easily understandable to the target population. Avoid complex jargon, technical terms, or ambiguous phrases that could lead to misinterpretation or confusion.
- **Provide Clear Instructions:** Clearly outline the purpose of the instrument and provide detailed instructions for completing each item or question. Specify response formats and specific guidelines for participants to follow.
- **Organize and Format:** Organize the instrument in a logical and sequential manner. Use headings, subheadings, and numbering to create a clear structure and facilitate ease of completion.

Relevance

- **Alignment with Research Objectives:** Ensure that each item or question in the instrument directly relates to the research objectives and the specific information needed to answer the research questions. Avoid including irrelevant or extraneous items that may unnecessarily burden participants.
- **Consider Participant Perspective:** Take into account the perspective of the participants when designing the instrument. Consider their experiences, knowledge, and cultural context to ensure that the instrument is relevant and meaningful to them.
- **Address Key Domains:** Identify the key domains or constructs to be measured and develop items that capture the relevant aspects of each domain. Ensure that the instrument provides comprehensive coverage of the intended constructs.

Validity

- **Content Validity:** Ensure that the instrument adequately covers all relevant aspects of the construct being measured. Conduct a thorough literature review and consult with experts in the field to ensure content validity.
- **Face Validity:** Review the instrument to ensure that the items appear to measure what they are intended to measure. Seek feedback from colleagues, experts, or representatives of the target population to assess face validity.
- **Construct Validity:** Establish the instrument's ability to measure the intended construct by exploring its relationship with other variables or constructs. Use established measures or validated instruments as benchmarks for comparison.
- **Criterion Validity:** Evaluate the instrument's ability to predict or correlate with an external criterion. Compare the instrument's results with established measures or objective criteria to assess criterion validity.

Measurement Scales

- **Select Appropriate Scales:** Choose measurement scales that are suitable for the research objectives and the nature of the variables being measured. Commonly used scales include Likert scales, visual analog scales, and semantic differential scales.
- **Consider Response Options:** Determine the response options that best capture participants' responses. These may include options such as numerical scales, multiple-choice options, or open-ended response formats. Consider providing an "other" or "not applicable" response option when necessary.
- **Balance Response Categories:** Ensure that response categories are balanced and cover the full range of possible responses. Avoid response options that are too skewed or overlapping, as they can introduce bias.

Pilot Testing

- **Conduct Pilot Testing:** Before finalizing the instrument, conduct pilot testing with a small group of participants who represent the target population. This helps identify any ambiguities, difficulties, or inconsistencies in the instrument and allows for refinements and improvements.
- **Assess Comprehension and Completion Time:** Evaluate participants' comprehension of the instrument and their ability to complete it within a reasonable time frame. Seek feedback on clarity, relevance, and overall usability.
- **Iterative Revision:** Based on the feedback from pilot testing, revise and refine the instrument iteratively. Consider the participants' suggestions, address any identified issues, and clarify ambiguous items or instructions.

Cultural and Linguistic Appropriateness

- **Consider Cultural Context:** Take into account cultural norms, beliefs, and values of the target population. Ensure that the instrument is sensitive to cultural differences and avoids any potentially offensive or biased content.
- **Translate and Adapt:** If the instrument will be used with a non-English-speaking population, translate it into the participants' language using established translation and back-translation procedures. Adapt the instrument to ensure cultural relevance and linguistic appropriateness.
- **Pilot Test with Diverse Participants:** Conduct pilot testing with diverse participants from different cultural backgrounds to assess the instrument's cultural and linguistic appropriateness. Seek input from bilingual or multicultural experts to enhance the instrument's relevance.

Conclusion

Designing data collection instruments is a critical step in clinical research. By adhering to key principles of clarity, relevance, and validity, researchers can develop instruments that effectively capture the required data. Considering measurement scales, conducting pilot testing, and ensuring cultural and linguistic appropriateness further enhance the instrument's quality. By designing instruments that are clear, relevant, and valid, researchers can collect accurate and meaningful data, contributing to the overall success of their research endeavors.

SECTION 3: DATA COLLECTION PROCEDURES

Data Collection Procedures

In this section, we will delve into the practical aspects of data collection procedures in clinical research. We will emphasize the importance of standardized protocols to ensure consistency, minimize potential bias, and enhance the reliability of study data. We will explore various strategies for participant recruitment and scheduling, data collection training for research staff, and maintaining data quality and integrity throughout the study. By implementing robust data collection procedures, researchers can minimize errors and enhance the overall quality of their research findings.

Standardized Protocols

- **Importance of Standardization:** Standardized protocols provide clear guidelines and instructions for data collection, ensuring consistency across different study sites, research staff, and time points. They minimize variations in data collection procedures and reduce the potential for bias, enhancing the validity and reliability of study findings.

- **Development of Data Collection Manuals:** Develop comprehensive data collection manuals that outline step-by-step procedures, instructions for administering questionnaires or conducting interviews, and guidelines for handling and documenting data. Ensure that the manuals are easily accessible to all research staff involved in data collection.

- **Training and Calibration:** Conduct thorough training sessions to familiarize research staff with the data collection protocols and procedures. Implement calibration exercises to ensure uniform understanding and adherence to the protocols. Regularly assess and reinforce training to maintain consistency throughout the study.

Participant Recruitment and Scheduling

- **Proactive Recruitment Strategies:** Develop proactive strategies for participant recruitment to ensure an adequate sample size. Utilize a variety of recruitment methods, such as healthcare providers, clinics, advertisements, online platforms, social media, and community engagement, to reach the target population effectively.

- **Efficient Scheduling:** Implement efficient scheduling procedures to optimize participant enrollment and data collection. Consider participants' preferences, availability, and logistical constraints when scheduling study visits or data collection sessions. Offer flexible options, such as evening or weekend appointments, to accommodate participants' needs.

Data Collection Training for Research Staff

- **Comprehensive Training Programs:** Design comprehensive training programs for research staff involved in data collection. Train them on study protocols, specific data collection instruments, ethical considerations, participant interactions, and data security and confidentiality. Provide ongoing support and resources to address questions and challenges that may arise during data collection.

- **Standardized Training Materials:** Develop standardized training materials, including training manuals, videos, and presentations, to ensure consistent training across different research staff. Incorporate interactive components, role-playing exercises, and case scenarios to enhance learning and comprehension.

- **Continuous Quality Improvement:** Implement mechanisms for ongoing evaluation and feedback to continuously improve data collection procedures. Regularly assess research staff's competence, adherence to protocols, and data quality. Provide constructive feedback, address any identified issues, and offer additional training or support as needed.

Data Quality and Integrity

- **Data Monitoring and Auditing:** Implement data monitoring and auditing procedures to ensure data quality and integrity. Regularly review data for accuracy, completeness, and adherence to standardized protocols. Conduct periodic data audits to identify any inconsistencies, errors, or deviations from protocols and take appropriate corrective actions.

- **Quality Control Checks:** Implement quality control checks during data collection, such as double data entry, verification of entered data against source documents, and range checks to identify outliers or data entry errors. Use data management software or systems to facilitate efficient data capture and reduce human error.

- **Data Security and Confidentiality:** Establish robust data security and confidentiality measures to protect participants' privacy and ensure compliance with ethical guidelines and regulations. Implement secure data storage and transmission methods, restrict access to authorized personnel only, and anonymize or encrypt data as necessary.

Conclusion

Implementing robust data collection procedures is crucial for ensuring the reliability and validity of study data in clinical research. Standardized protocols, participant recruitment and scheduling strategies, comprehensive training for research staff, and data quality and integrity measures all contribute to the overall success of data collection. By adhering to these procedures and addressing potential challenges proactively, researchers can minimize errors, enhance the quality of their data, and ultimately strengthen the validity and reliability of their research findings.

SECTION 4: ELECTRONIC DATA CAPTURE AND TECHNOLOGY IN DATA COLLECTION

Electronic Data Capture and Technology in Data Collection

In this section, we will focus on the use of electronic data capture (EDC) systems and technology in data collection. We will explore the benefits of EDC systems, including improved data accuracy, efficiency, and security. Additionally, we will discuss different EDC platforms and tools available for data collection, management, and analysis. Readers will gain insights into the considerations involved in selecting and implementing EDC systems, including data privacy and security, user-friendliness, and integration with other research systems. Furthermore, we will discuss the use of technology, such as mobile applications and wearable devices, in collecting real-time data and enhancing participant engagement.

Benefits of EDC Systems

- **Improved Data Accuracy:** EDC systems reduce the reliance on manual data entry, minimizing the risk of transcription errors and data entry mistakes. The use of electronic forms and data validation checks enhances data accuracy and completeness, leading to more reliable research findings.

- **Increased Efficiency:** EDC systems streamline data collection processes by automating data capture, eliminating the need for paper-based forms and manual data entry. This improves the efficiency of data collection, reduces the time required for data management, and accelerates the overall research timeline.

- **Enhanced Data Security:** EDC systems offer robust data security measures, including role-based access control, data encryption, and audit trails. These features ensure the confidentiality and integrity of participant data, complying with ethical and regulatory requirements.

EDC Platforms and Tools

- **Selection Considerations:** When selecting an EDC platform or tool, researchers should consider factors such as data privacy and security features, user-friendliness, system scalability, and compatibility with existing research infrastructure. Evaluate the platform's ability to handle different data types, integrate with other systems (e.g., electronic health records), and support data export and analysis.

- **Common EDC Platforms:** Some commonly used EDC platforms include REDCap (Research Electronic Data Capture), OpenClinica, Medidata Rave, and Oracle Clinical. These platforms offer a range of features for data collection, management, and analysis, allowing researchers to customize forms, automate data validation, and generate reports.

- **Electronic Case Report Forms (eCRFs):** Electronic case report forms are the electronic equivalents of paper-based data collection forms. They can be designed using EDC platforms, providing structured data entry fields, validation checks, and skip patterns. eCRFs improve data accuracy and facilitate efficient data collection.

Technology in Data Collection

- **Mobile Applications:** Mobile applications allow for data collection using smartphones or tablets, providing flexibility and convenience. They enable real-time data capture, including multimedia elements (e.g., photos, videos, and audio recordings), location data, and participant-reported outcomes. Mobile apps can improve participant engagement and compliance with study protocols.

- **Wearable Devices:** Wearable devices, such as fitness trackers or smartwatches, can collect objective data in real-time, including physical activity levels, heart rate, sleep patterns, and other physiological measurements. These devices offer the advantage of continuous data monitoring without relying on participant self-reporting.

- **Remote Data Collection:** Technology enables remote data collection, reducing the need for participants to visit study sites. Telehealth consultations, remote monitoring, and electronic diaries empower participants to provide data from the comfort of their homes, improving convenience, reducing participant burden, and increasing study enrollment.

Data Privacy and Security

- **Compliance with Regulations:** When using EDC systems and technology, researchers must adhere to data privacy and security regulations, such as the General Data Protection Regulation (GDPR) or Health Insurance Portability and Accountability Act (HIPAA). Implement appropriate measures to protect participant privacy, including data encryption, secure data storage, access controls, and informed consent for data sharing.

- **Data Backup and Disaster Recovery:** Ensure regular data backups and implement robust disaster recovery plans to prevent data loss. Consider off-site data storage options and redundancy measures to ensure data availability and integrity.

- **Participant Data Ownership:** Clearly communicate participant data ownership and rights during the informed consent process. Inform participants about how their data will be stored, used, and shared, and provide options for data withdrawal or participant data anonymization.

Conclusion

The utilization of electronic data capture systems and technology in data collection brings numerous benefits to clinical research, including improved data accuracy, efficiency, and security. Researchers should carefully consider the selection of EDC platforms and tools, considering data privacy, user-friendliness, and system integration. Additionally, the integration of technology, such as mobile applications and wearable devices, allows for real-time data collection and enhances participant engagement. However, researchers must ensure compliance with data privacy and security regulations, prioritize data backup and disaster recovery, and address participant data ownership concerns. By harnessing the potential of EDC systems and technology, researchers can optimize data collection processes, improve research outcomes, and advance the field of clinical research.

SECTION 5: DATA QUALITY ASSURANCE AND MONITORING

Data Quality Assurance and Monitoring

In this section, we will delve into data quality assurance and monitoring in clinical research. We will discuss strategies for ensuring data accuracy, completeness, and consistency, highlighting the importance of data quality control processes. Additionally, we will explore various techniques such as data validation checks, double data entry, and regular data monitoring to maintain the integrity of study data. Readers will gain insights into the role of data monitoring committees in overseeing data quality and receive practical guidance on implementing data quality assurance measures.

Data Quality Assurance

- **Data Validation Checks:** Implementing data validation checks is crucial to ensure data accuracy and completeness. These checks include range checks, logical checks, and consistency checks that flag any potential errors or inconsistencies in the data. By setting validation rules and conducting systematic checks, researchers can identify and correct data entry errors promptly.

- **Double Data Entry:** Double data entry involves independent entry of data by two different individuals, followed by a comparison of the two entries to identify discrepancies. This process helps identify data entry errors, reducing the risk of incorrect or incomplete data.

- **Standard Operating Procedures (SOPs):** Developing SOPs for data collection, data entry, and data management processes establishes standardized practices and guidelines for research staff. SOPs outline the steps to be followed, data quality control procedures, and roles and responsibilities of team members. Clear and well-documented SOPs promote consistency and enhance data quality.

Data Monitoring

- **Regular Data Monitoring:** Regular data monitoring involves ongoing review and assessment of study data to identify outliers, missing data, or any potential data quality issues. Researchers should establish a schedule for data monitoring and conduct routine checks to ensure data completeness, accuracy, and adherence to study protocols.

- **Data Monitoring Committees (DMCs):** DMCs play a crucial role in overseeing data quality and integrity in clinical research, particularly in large, multicenter trials. These independent committees review and analyze study data, assess participant safety, and provide recommendations on data quality and trial conduct. DMCs help ensure unbiased monitoring and safeguard the reliability and validity of study findings.

Data Quality Control Processes

- **Training and Standardization:** Thorough training of research staff involved in data collection and data entry is essential to minimize errors. Providing clear instructions and conducting training sessions on data collection tools, data entry systems, and data quality control procedures promote consistency and improve data accuracy.

- **Data Cleaning:** Data cleaning involves identifying and resolving data discrepancies, outliers, and missing values. Researchers should establish a systematic process for data cleaning, including procedures for resolving data queries and conducting data reconciliation. Regular data cleaning procedures improve data quality and ensure the reliability of research findings.

- **Data Audits:** Periodic data audits, conducted either internally or by external auditors, can help assess the overall quality and integrity of study data. Audits involve a comprehensive review of data collection processes, data management procedures, and adherence to data quality control measures. The findings from data audits inform researchers about areas for improvement and reinforce data quality assurance practices.

Conclusion

Data quality assurance and monitoring are essential components of clinical research to ensure data accuracy, completeness, and consistency. By implementing data validation checks, double data entry, and regular data monitoring, researchers can minimize errors and maintain the integrity of study data. Establishing data quality control processes, training research staff, and involving data monitoring committees contribute to the overall reliability and validity of research findings. By prioritizing data quality assurance measures, researchers can enhance the trustworthiness of their research outcomes and contribute to the advancement of clinical knowledge.

SECTION 6: DATA PRIVACY, CONFIDENTIALITY, AND SECURITY

Data Privacy, Confidentiality, and Security

In this section, we will focus on the critical aspects of data privacy, confidentiality, and security in clinical research. We will explore the ethical and legal considerations surrounding the protection of participant data and discuss strategies for ensuring data privacy and confidentiality. Additionally, we will highlight key concepts such as data anonymization, secure data storage, and access control. Readers will gain insights into the regulatory frameworks governing data privacy and security, including HIPAA and GDPR. We will emphasize the importance of adhering to ethical guidelines and maintaining the trust and confidentiality of research participants.

Ethical Considerations

- **Informed Consent:** Obtaining informed consent from research participants includes informing them about the privacy and confidentiality measures in place to protect their data. Participants should be made aware of how their data will be collected, stored, accessed, and shared, and any associated risks or limitations. Researchers must ensure that participants fully understand and voluntarily consent to the use of their data.

- **Anonymization and De-identification:** Data anonymization is a key strategy to protect participant privacy. By removing or encrypting personally identifiable information (PII) from data sets, researchers can minimize the risk of re-identification. De-identification techniques, such as removing direct identifiers or modifying data elements, further protect participant confidentiality.

- **Data Access and Handling:** Researchers should establish protocols for data access and handling to limit unauthorized access and maintain confidentiality. This includes assigning access privileges based on the need-to-know principle, implementing secure authentication measures, and monitoring data access logs. Researchers should also ensure secure transmission and storage of data to prevent unauthorized disclosure or breaches.

Legal Frameworks and Regulations

- **HIPAA (Health Insurance Portability and Accountability Act):** HIPAA regulations in the United States provide privacy and security protections for protected health information (PHI). Researchers conducting studies involving PHI must comply with HIPAA requirements, including obtaining participant authorization, implementing safeguards to protect PHI, and ensuring data breaches are promptly reported.

- **GDPR (General Data Protection Regulation):** GDPR applies to studies conducted within the European Union (EU) or involving EU citizens' personal data. Researchers must obtain explicit consent for data collection, clearly state the purpose of data processing, and implement appropriate technical and organizational measures to ensure data security and confidentiality.

Data Security Measures

- **Secure Data Storage:** Researchers should utilize secure data storage systems with appropriate access controls, encryption, and firewalls to protect against unauthorized access. Data should be stored in controlled environments, such as secure servers or cloud-based platforms, with regular backups and disaster recovery plans in place.

- **Access Control:** Implementing access controls, such as unique user logins and role-based permissions, helps restrict data access to authorized individuals only. Regularly review and update access permissions to ensure that access privileges are granted based on the principle of least privilege.

- **Data Transfer and Sharing:** When sharing data with collaborators or third parties, researchers must use secure methods, such as encrypted file transfers or secure data-sharing platforms. Data transfer agreements should be established to outline responsibilities and protect participant privacy during data sharing.

Conclusion

Data privacy, confidentiality, and security are paramount in clinical research to protect the rights and privacy of research participants. By adhering to ethical guidelines and regulatory frameworks, such as HIPAA and GDPR, researchers can ensure the privacy and confidentiality of participant data. Strategies such as data anonymization, secure data storage, and access control help safeguard participant information. It is essential for researchers to maintain the trust of participants by communicating privacy and security measures clearly and obtaining informed consent. By implementing robust data privacy and security practices, researchers can uphold ethical standards, protect participant confidentiality, and contribute to the integrity and credibility of clinical research.

SECTION 7: DATA MANAGEMENT AND DOCUMENTATION

Data Management and Documentation

In this final section, we will delve into the crucial aspects of data management and documentation in clinical research. Effective data management ensures that study data are organized, well-documented, and readily accessible. We will discuss strategies for organizing and documenting study data, including the use of data dictionaries, coding systems, and data documentation templates. Additionally, we will emphasize the importance of data archiving and retention for long-term accessibility and reproducibility of research findings. By implementing best practices in data management and documentation, researchers can enhance the integrity and efficiency of their studies while ensuring compliance with regulatory requirements.

Data Management Strategies

- **Data Organization:** Researchers should establish a structured system for organizing study data, including creating a clear folder structure, file naming conventions, and version control. This helps ensure that data files are organized in a logical and consistent manner, making it easier to locate and retrieve specific datasets or documents.

- **Data Dictionaries:** Developing data dictionaries that provide a comprehensive description of variables, their definitions, and any coding or formatting rules is essential. Data dictionaries serve as a reference guide, ensuring consistency in data entry and facilitating data analysis.

- **Coding Systems:** Researchers may utilize coding systems, such as standardized medical coding schemes (e.g., ICD-10 or SNOMED-CT) or specific coding systems for qualitative data (e.g., thematic coding), to categorize and analyze data consistently. Clear guidelines should be established for coding practices to enhance accuracy and reliability.

Data Documentation

- **Data Collection Forms:** Researchers should create well-designed data collection forms that capture all necessary variables and ensure accurate data entry. These forms should be standardized and pre-tested to minimize errors and inconsistencies during data collection.

- **Data Documentation Templates:** Researchers may develop templates for data documentation, including data entry logs, data cleaning and validation procedures, and data audit trails. These templates help document the data management process and ensure transparency and accountability.

- **Metadata Documentation:** Metadata, which describes the context, structure, and content of the data, is crucial for understanding and interpreting the dataset. Researchers should document metadata, including study design, variables, units of measurement, and data collection methods, to provide comprehensive information for future users.

Data Archiving and Retention

- **Data Archiving:** Researchers should establish protocols for data archiving, ensuring that study data are securely stored and preserved for long-term accessibility. Archiving may involve storing data on secure servers, cloud-based platforms, or dedicated data repositories. Compliance with regulatory requirements and ethical considerations, such as participant privacy and data security, should be ensured during data archiving.

- **Data Retention:** It is important to determine the appropriate duration for retaining study data based on institutional policies, funding agency requirements, and regulatory guidelines. Retaining data for an adequate period allows for data verification, potential future analysis, and reproducibility of research findings.

Conclusion

Effective data management and documentation are crucial for the success and integrity of clinical research. By implementing strategies such as data organization, data dictionaries, and coding systems, researchers can ensure that study data are well-organized, consistent, and easily retrievable. Proper documentation, including data collection forms, data documentation templates, and metadata, enhances transparency and facilitates data interpretation. Furthermore, data archiving and retention practices enable long-term accessibility and reproducibility of research findings. By adhering to best practices in data management and documentation, researchers can promote data integrity, facilitate collaboration, and comply with regulatory requirements, ultimately advancing the field of clinical research.

By the end of Chapter 6, readers will have a comprehensive understanding of data collection and management in clinical research. They will grasp the importance of collecting high-quality data, designing data collection instruments, implementing data collection procedures, ensuring data privacy and security, and practicing effective data management and documentation. This knowledge will empower readers to implement robust data collection and management practices, leading to reliable and meaningful research outcomes.

7

DATA ANALYSIS AND INTERPRETATION IN CLINICAL RESEARCH

CHAPTER VII

DATA ANALYSIS AND INTERPRETATION IN CLINICAL RESEARCH

Chapter 7 delves into the critical aspects of data analysis and interpretation in clinical research. The focus is on providing researchers with a thorough understanding of the fundamental principles, techniques, and tools necessary to analyze and interpret research data effectively. This chapter covers various types of data analysis methods, including descriptive statistics, inferential statistics, and qualitative data analysis, emphasizing their application in clinical research. It also explores advanced data analysis techniques and the use of statistical software. By the end of this chapter, readers will be equipped with the knowledge and skills to perform robust data analysis, draw meaningful conclusions, and effectively communicate their research findings to the scientific community and other stakeholders.

SECTION 1: INTRODUCTION TO DATA ANALYSIS

Introduction to Data Analysis

Data analysis in clinical research is a critical process that transforms collected data into meaningful insights, addressing research questions and hypotheses. This section introduces the fundamental concepts and techniques of data analysis, emphasizing the importance of accuracy, consistency, and methodological rigor. Readers will explore various types of data analysis methods, including descriptive and inferential statistics, qualitative data analysis, and exploratory data analysis. By understanding these principles, researchers can effectively analyze their data, draw reliable conclusions, and contribute valuable knowledge to the field of clinical research.

Role of Data Analysis in Clinical Research

Data analysis is a vital component of the research process in clinical studies. It involves transforming raw data into meaningful information, identifying patterns and trends, and drawing inferences and conclusions. The primary objectives of data analysis are to:

- **Explore and Summarize Data:** Data analysis allows researchers to examine the characteristics of their data, identify patterns, and summarize key findings. It provides a comprehensive understanding of the variables and their distributions, enabling researchers to describe and present the data effectively.

- **Address Research Questions:** Data analysis is crucial for addressing research questions and hypotheses. By applying appropriate statistical methods and analytical techniques, researchers can examine relationships, test hypotheses, and draw conclusions based on the evidence provided by the data.

- **Extract Insights and Make Inferences:** Data analysis enables researchers to derive insights and make inferences about the population under study. By extrapolating from a representative sample, researchers can generalize their findings to the broader target population, enhancing the applicability and generalizability of their results.

Importance of Data Cleaning and Preparation

Before proceeding with data analysis, researchers must perform data cleaning and preparation to ensure the accuracy, completeness, and quality of the data. This process involves:

- **Data Validation and Verification:** Researchers should carefully validate and verify the accuracy and completeness of the collected data. This includes checking for missing values, outliers, and inconsistencies, and addressing any data entry errors or discrepancies.

- **Data Transformation and Standardization:** Data may require transformation and standardization to ensure compatibility and comparability. This may involve converting data formats, recoding variables, and standardizing units of measurement.

- **Dealing with Missing Data:** Researchers must handle missing data appropriately to minimize bias and ensure the integrity of the analysis. This may involve imputing missing values using appropriate statistical methods or considering the implications of missing data in the interpretation of results.

Steps in the Data Analysis Process

The data analysis process typically involves the following steps:

- **Data Exploration:** Researchers explore the data by examining summary statistics, frequency distributions, and visualizations to gain a preliminary understanding of the variables and identify any patterns or outliers.

- **Data Transformation:** If necessary, researchers may perform data transformation to meet the assumptions of the chosen statistical methods. This may include logarithmic transformations, standardization, or normalization.

- **Statistical Analysis:** Researchers select and apply appropriate statistical methods based on the research questions, study design, and nature of the data. This may involve descriptive statistics, inferential statistics, qualitative data analysis, or exploratory data analysis techniques.

- **Interpretation of Results:** Researchers interpret the results obtained from the analysis, considering the statistical significance, effect sizes, and practical implications. They draw conclusions based on the evidence provided by the data and relate them to the research objectives and existing literature.

Types of Data Analysis Methods

Clinical research employs various data analysis methods, including:

- **Descriptive Statistics:** Descriptive statistics summarize and describe the main characteristics of the data, such as measures of central tendency (mean, median) and variability (standard deviation, range).

- **Inferential Statistics:** Inferential statistics allow researchers to make inferences and draw conclusions about the population based on sample data. These methods include hypothesis testing, confidence intervals, and regression analysis.

- **Qualitative Data Analysis:** Qualitative data analysis involves systematically analyzing non-numerical data, such as interview transcripts or observational notes. It employs techniques like thematic analysis, content analysis, or grounded theory to identify patterns and themes.

Conclusion

Data analysis is a crucial component of clinical research, enabling researchers to extract meaningful insights, address research questions, and make evidence-based decisions. By understanding the role of data analysis, researchers can effectively analyze their data, draw accurate conclusions, and contribute to advancements in healthcare practices. Furthermore, proper data cleaning and preparation are essential to ensure the quality and reliability of the analysis. By employing appropriate data analysis methods, such as descriptive statistics, inferential statistics, qualitative data analysis, and exploratory data analysis, researchers can uncover valuable information and contribute to the body of knowledge in clinical research.

SECTION 2: DESCRIPTIVE STATISTICS

Descriptive Statistics

Descriptive statistics play a crucial role in summarizing and presenting data in a meaningful and concise manner. In this section, we will explore measures of central tendency and measures of variability, which provide insights into the characteristics of a dataset. We will discuss the calculation and interpretation of these descriptive statistics, along with their application in clinical research. Additionally, we will explore graphical representations, such as histograms, box plots, and scatter plots, which visually represent data distribution and relationships. By understanding and utilizing descriptive statistics, researchers can effectively summarize and interpret their data, enhancing the understanding of their research findings.

Measures of Central Tendency

Measures of central tendency describe the typical or central value of a dataset. The three commonly used measures of central tendency are:

- **Mean:** The mean is calculated by summing all the values in a dataset and dividing it by the total number of observations. It represents the average value of the data. The mean is sensitive to extreme values, making it important to consider the presence of outliers.

- **Median:** The median is the middle value in a dataset when it is arranged in ascending or descending order. It is not affected by extreme values and provides a robust measure of central tendency. The median is particularly useful when dealing with skewed distributions or ordinal data.

- **Mode:** The mode represents the most frequently occurring value in a dataset. It is applicable for both numerical and categorical data. In cases where multiple values have the same highest frequency, the dataset is considered multimodal.

These measures of central tendency provide a summary of where the data is centered, allowing researchers to gain insights into the typical values of the variables under study.

Measures of Variability

Measures of variability describe the spread or dispersion of data points around the measures of central tendency. They provide information about the range and distribution of the dataset. The three commonly used measures of variability are:

- **Range:** The range is calculated by subtracting the minimum value from the maximum value in a dataset. It provides a simple and quick assessment of the spread of data. However, it is sensitive to outliers and may not provide a robust measure of variability.

- **Standard Deviation:** The standard deviation measures the average distance of data points from the mean. It quantifies the spread of data and indicates the typical deviation from the mean. A higher standard deviation suggests a greater dispersion of data points.

- **Variance:** The variance is the average of the squared deviations from the mean. It provides a measure of variability by considering the dispersion of each data point from the mean. The variance is commonly used in statistical calculations, and its square root is equal to the standard deviation.

These measures of variability provide insights into the spread of data, allowing researchers to understand the variability and dispersion of their variables of interest.

Graphical Representations

Graphical representations are valuable tools for visualizing data distribution and relationships. They help researchers gain a better understanding of the dataset and communicate findings effectively. Some commonly used graphical representations include:

- **Histograms:** Histograms display the frequency distribution of numerical data. They present data in bins or intervals on the x-axis and the frequency or proportion of observations on the y-axis. Histograms provide insights into the shape, central tendency, and spread of the data.

- **Box Plots:** Box plots, also known as box-and-whisker plots, provide a visual summary of the distribution of data. They display the median, quartiles, and outliers in a dataset. Box plots are particularly useful for comparing groups or variables and identifying skewness or outliers.

- **Scatter Plots:** Scatter plots depict the relationship between two continuous variables. Each data point represents a pair of values, and the plot shows how they are distributed. Scatter plots help identify patterns, trends, and the strength of the relationship between variables.

These graphical representations enhance the understanding of data patterns, allowing researchers to interpret relationships and detect outliers or unusual observations.

Conclusion

Descriptive statistics, including measures of central tendency and measures of variability, provide valuable insights into the characteristics of a dataset in clinical research. By calculating and interpreting these statistics, researchers can summarize and understand the central values and spread of their variables of interest. Additionally, graphical representations, such as histograms, box plots, and scatter plots, offer visual tools to explore data distribution and relationships. By combining descriptive statistics and graphical representations, researchers can effectively communicate their findings and gain a deeper understanding of the data in their clinical research studies.

SECTION 3: INFERENTIAL STATISTICS

Inferential Statistics

Inferential statistics play a crucial role in clinical research by allowing researchers to draw conclusions and make inferences about a population based on sample data. In this section, we will explore the concept of hypothesis testing, which involves selecting appropriate statistical tests, interpreting the results, and determining the statistical significance of observed differences or associations. We will discuss common statistical tests used in clinical research, such as t-tests, chi-square tests, and ANOVA. Additionally, we will delve into important concepts such as p-values, confidence intervals, and statistical power. It is essential to understand the limitations and assumptions of inferential statistics and interpret results within the context of clinical relevance.

Hypothesis Testing

Hypothesis testing is a fundamental concept in inferential statistics. It involves formulating a null hypothesis (Ho) that assumes no significant difference or association between variables in the population, and an alternative hypothesis (Ha) that suggests the presence of a significant difference or association. The goal is to gather evidence from the sample data to either reject the null hypothesis in favor of the alternative hypothesis or fail to reject the null hypothesis.

Selecting Appropriate Statistical Tests

Selecting the appropriate statistical test depends on the research question, study design, and nature of the variables being analyzed. Here are a few commonly used statistical tests in clinical research:

- **t-tests:** t-tests are used to compare means between two groups or assess the difference between a sample mean and a known population mean. They can be independent (e.g., comparing means between two independent groups) or paired (e.g., comparing means within the same group before and after an intervention).

- **Chi-square tests:** Chi-square tests are used to assess the association between categorical variables. They can determine if there is a significant difference between observed and expected frequencies in a contingency table.

- **ANOVA (Analysis of Variance):** ANOVA is used to compare means across three or more groups. It helps determine if there are significant differences between the means of multiple groups.

Selecting the appropriate statistical test is crucial to ensure that the analysis aligns with the research question and data characteristics.

Interpreting Results

Interpreting the results of statistical tests involves assessing the statistical significance of the findings. Key concepts in interpreting results include:

- **p-values:** The p-value represents the probability of obtaining the observed results or more extreme results if the null hypothesis is true. A p-value less than a predetermined significance level (typically 0.05) indicates statistical significance, suggesting that the observed results are unlikely to have occurred by chance.

- **Confidence Intervals:** Confidence intervals provide a range of values within which the true population parameter is likely to lie. They offer information on the precision of the estimate and can help assess the clinical relevance of the findings.

- **Statistical Power:** Statistical power is the probability of correctly detecting a true effect or association. It depends on factors such as sample size, effect size, and alpha level. Higher statistical power increases the likelihood of identifying significant findings when they exist.

Limitations and Assumptions

It is essential to recognize the limitations and assumptions of inferential statistics. Common limitations include the reliance on sample data, which may not fully represent the population, and the assumption of random sampling and independence. Violations of these assumptions can impact the validity of the results.

Interpreting the results of statistical tests should always be done in the context of clinical relevance. Statistical significance does not necessarily imply clinical significance, and it is crucial to consider the magnitude of the effect, the clinical relevance of the findings, and the potential impact on patient care.

Conclusion

Inferential statistics are a powerful tool in clinical research, enabling researchers to draw conclusions and make inferences about a population based on sample data. Hypothesis testing, selecting appropriate statistical tests, and interpreting the results are essential components of inferential statistics. Understanding concepts such as p-values, confidence intervals, and statistical power is crucial for accurate interpretation. However, it is important to recognize the limitations and assumptions of inferential statistics and interpret results within the context of clinical relevance. By applying inferential statistics appropriately and critically interpreting the findings, researchers can contribute valuable insights to the field of clinical research.

SECTION 4: QUALITATIVE DATA ANALYSIS

Qualitative Data Analysis

Qualitative data analysis is a valuable approach used in clinical research to explore and understand the complexities of human experiences, perceptions, and behaviors. In this section, we will delve into qualitative data analysis methods, including thematic analysis, content analysis, and grounded theory. We will discuss the steps involved in qualitative data analysis and strategies for enhancing rigor and trustworthiness. By systematically analyzing textual or narrative data, researchers can identify patterns, themes, and insights that contribute to a deeper understanding of the research phenomenon.

Approaches to Qualitative Data Analysis

- **Thematic Analysis:** Thematic analysis involves identifying and organizing themes or patterns in the data. Researchers systematically examine the data to identify recurring concepts, ideas, or experiences and develop thematic categories. This approach allows for a comprehensive exploration of the data and can uncover both anticipated and unexpected themes.

- **Content Analysis:** Content analysis involves analyzing the content of qualitative data, such as interviews, focus groups, or written documents. Researchers identify and code specific words, phrases, or concepts to categorize the data and identify patterns or trends. This approach is particularly useful when exploring large datasets or when the research objective is to quantify qualitative data.

- **Grounded Theory:** Grounded theory aims to develop a theoretical understanding based on the data. It involves systematically analyzing the data and developing concepts or theories that emerge directly from the data itself, rather than starting with preconceived theories. Researchers iteratively collect, code, and analyze data to refine and develop theoretical frameworks.

Steps in Qualitative Data Analysis

- **Data Coding:** Data coding involves systematically assigning labels or codes to segments of data that represent different concepts, themes, or categories. Researchers read and reread the data, annotate relevant sections, and develop a coding scheme. Coding can be deductive (using preexisting concepts or theories) or inductive (allowing concepts to emerge from the data).

- **Categorization and Theme Development:** Once data coding is complete, researchers organize the codes into meaningful categories or themes. This process involves grouping related codes, identifying overarching patterns, and developing a coherent narrative.

- **Interpretation and Analysis:** Researchers interpret the categorized data and analyze the relationships between themes. They explore connections, contradictions, and variations within the data to generate insights and develop a comprehensive understanding of the research phenomenon.

Enhancing Rigor and Trustworthiness

- **Intercoder Reliability:** Intercoder reliability involves multiple researchers independently coding a subset of the data and assessing the agreement between their codes. Establishing high intercoder reliability enhances the rigor and credibility of the analysis by ensuring consistency and reducing individual bias.

- **Member Checking:** Member checking involves sharing the analyzed findings with participants to verify the accuracy and resonance of the interpretations. This process allows participants to provide feedback, clarify any misunderstandings, or offer additional insights, further enhancing the trustworthiness of the analysis.

Considerations in Qualitative Data Analysis

- **Reflexivity:** Researchers should be aware of their own biases, assumptions, and preconceptions that might influence the data analysis process. Reflexivity involves reflecting on and documenting the researcher's perspectives and experiences to ensure transparency and minimize the impact of personal biases.

- **Saturation:** Saturation refers to the point in qualitative data analysis when new data no longer reveals additional information or themes. Researchers aim to achieve data saturation to ensure that the analysis captures the breadth and depth of the research phenomenon.

- **Triangulation:** Triangulation involves using multiple sources of data or multiple researchers to analyze the data. By incorporating diverse perspectives and data sources, researchers can strengthen the validity and reliability of the analysis.

Conclusion

Qualitative data analysis provides a rich and nuanced understanding of human experiences and phenomena in clinical research. Through approaches such as thematic analysis, content analysis, and grounded theory, researchers can systematically analyze and interpret textual or narrative data. By following rigorous procedures, such as data coding, categorization, and interpretation, researchers can derive meaningful insights and contribute to the body of knowledge in clinical research. Additionally, strategies such as intercoder reliability and member checking enhance the trustworthiness and credibility of the analysis. Qualitative data analysis complements quantitative methods, providing a comprehensive understanding of complex research topics and fostering a holistic approach to evidence-based practice.

SECTION 5: STATISTICAL SOFTWARE AND TOOLS

Statistical Software and Tools

This section will focus on statistical software and tools commonly used in clinical research data analysis. We will discuss popular statistical software packages such as SPSS, SAS, R, and STATA, and explore their features and capabilities. Readers will gain insights into data manipulation, data analysis, and graphical presentation using these software tools. We will provide practical tips and resources for learning and utilizing statistical software effectively.

SPSS (Statistical Package for the Social Sciences)

SPSS is a widely used statistical software package known for its user-friendly interface and extensive capabilities for data manipulation, statistical analysis, and graphical presentation. Key features include:

- **Data Management:** SPSS allows researchers to import, clean, and manipulate data efficiently. It provides tools for data transformation, merging datasets, handling missing values, and recoding variables.

- **Statistical Analysis:** SPSS offers a wide range of statistical procedures, including descriptive statistics, inferential tests (e.g., t-tests, ANOVA, regression analysis), nonparametric tests, survival analysis, and factor analysis. It provides both menu-driven and syntax-based approaches for conducting analyses.

- **Graphical Presentation:** SPSS enables researchers to create various types of graphs and charts, such as histograms, bar charts, scatter plots, and box plots. It offers customization options for visualizing data.

Practical Tips and Resources for SPSS

- Familiarize yourself with the basic functions and features of SPSS through online tutorials and user guides.
- Utilize the syntax mode for reproducibility and efficient data analysis.
- Join SPSS user forums or communities to seek help, share experiences, and learn from other users.
- Take advantage of SPSS' extensive online resources, including official documentation, sample datasets, and tutorials.

SAS (Statistical Analysis System)

SAS is a comprehensive statistical software suite widely used in clinical research. It offers a robust set of tools for data management, statistical analysis, and reporting. Key features of SAS include:

- **Data Management:** SAS provides powerful data manipulation capabilities, allowing researchers to handle large datasets . It offers data cleaning, merging, and transformation techniques for preparing data for analysis.

- **Statistical Analysis:** SAS offers a broad range of statistical procedures, including parametric and nonparametric tests, regression analysis, survival analysis, mixed models, and longitudinal data analysis. SAS also provides advanced procedures for complex statistical modeling.

- **Programming Flexibility:** SAS utilizes a programming language called SAS programming. It allows researchers to write custom programs, automate analyses, and create reproducible workflows. SAS also supports a point-and-click interface for those who prefer a graphical approach.

Practical Tips and Resources for SAS

- Attend SAS training courses or workshops to gain proficiency in using the software effectively.
- Make use of the extensive SAS documentation, including user guides, technical papers, and support resources.
- Engage with the SAS user community through forums, conferences, and user groups to learn from other users.
- Take advantage of SAS online learning resources, including video tutorials, webinars, and sample code repositories.

R

R is a free and open-source programming language and software environment for statistical computing and graphics. It has gained popularity among researchers due to its versatility, extensive statistical libraries, and active user community. Key features of R include:

- **Data Manipulation and Analysis:** R provides powerful functions and packages for data manipulation, transformation, and analysis. It offers a wide range of statistical techniques, from basic descriptive statistics to advanced modeling and machine learning algorithms.

- **Graphics and Data Visualization:** R excels in data visualization, offering numerous packages for creating publication-quality plots, charts, and graphs. Researchers can customize visuals to effectively communicate findings.

- **Extensibility and Collaboration:** R's open-source nature allows users to create and share their packages, expanding the functionality and capabilities of the software. R also promotes collaboration through the sharing of code, data, and analysis pipelines.

Practical Tips and Resources for R

- Start with introductory R tutorials and online courses to grasp the basics of R programming and data analysis.
- Explore the vast collection of R packages available on the Comprehensive R Archive Network (CRAN) and other repositories to leverage existing functions and methods.
- Join R user groups, forums, and online communities to interact with fellow researchers and seek guidance.

STATA

STATA is a statistical software package widely used in various fields, including clinical research. It offers a comprehensive set of tools for data management, statistical analysis, and visualization. Key features of STATA include:

- **Data Management:** STATA provides a range of features for data manipulation and cleaning. It allows researchers to handle complex data structures, merge datasets, reshape data, and handle missing values efficiently.

- **Statistical Analysis:** STATA offers a wide range of statistical procedures, including regression analysis, survival analysis, panel data analysis, and categorical data analysis. It also provides specialized modules for econometrics and biostatistics.

- **Publication-Ready Graphics:** STATA allows researchers to create high-quality graphics for data visualization and presentation. It offers customization options to generate publication-ready graphs and charts.

Practical Tips and Resources for STATA

- Familiarize yourself with STATA's command syntax and programming language to perform advanced data management and analysis tasks.
- Take advantage of STATA's extensive documentation, including user guides, manuals, and tutorials.
- Explore STATA's user community resources, such as forums and mailing lists, to seek assistance and learn from experienced users.

Conclusion

In this section, we provided a comprehensive understanding of popular statistical software packages used in clinical research data analysis, including SPSS, SAS, R, and STATA. We explored their features, capabilities, and provided practical tips and resources for effective utilization. Researchers should select the software package that aligns with their specific research needs, data complexity, and analysis requirements. Familiarity with these statistical software tools will enable researchers to efficiently analyze their data, draw meaningful conclusions, and contribute to evidence-based practice in the field of clinical research.

SECTION 6: INTERPRETING AND REPORTING STUDY FINDINGS

Interpreting Study Findings

Interpreting study findings involves analyzing the results of data analysis considering the research objectives and relevant literature. The following steps can help researchers effectively interpret their findings:

1. **Understand the Research Question:** Start by revisiting the research question and objectives to ensure a clear understanding of what the study aimed to investigate.
2. **Examine Descriptive Statistics:** Review descriptive statistics, such as means, medians, and proportions, to understand the characteristics and distribution of the variables under investigation.
3. **Consider Inferential Statistics:** If inferential statistics were used, analyze the results of statistical tests to determine if the observed differences or associations are statistically significant. Assess the magnitude and direction of the effects or relationships.
4. **Compare with Existing Literature:** Compare the study findings with existing literature to determine if they align with previous research or contribute new insights. Identify consistencies, inconsistencies, or gaps in knowledge.
5. **Explore Clinical Relevance:** Interpret the results in the context of clinical relevance. Consider the impact of the findings on patient care, treatment guidelines, or future research directions.

Reporting Study Findings

Reporting study findings involves effectively communicating research results to various audiences. Clear and concise reporting is essential to ensure the dissemination and understanding of research outcomes. Consider the following aspects when reporting study findings:

- **Transparent Reporting:** Provide a clear and detailed account of the study design, methods, and analysis procedures. Transparent reporting facilitates the reproducibility and credibility of the research.
- **Structure the Report:** Organize the report using a logical structure, including an introduction, methods, results, and discussion sections. Clearly state the research question, objectives, and hypothesis (if applicable) at the outset.
- **Address Study Limitations:** Acknowledge and discuss the limitations of the study, including potential biases, confounding factors, and sample size limitations. Addressing limitations demonstrates the researcher's awareness of potential sources of error and increases the robustness of the study's interpretation.
- **Use Visual Presentations:** Utilize appropriate visual presentations, such as tables, graphs, and figures, to enhance the clarity and understanding of the findings. Ensure that visuals are labeled, easy to interpret, and aligned with the main message.
- **Write Scientific Manuscripts:** When preparing scientific manuscripts, adhere to the guidelines of the target journal. Clearly present the research question, methods, results, and discussion. Use a concise and precise writing style, and include relevant references to support the interpretation of findings.
- **Disseminate Results to Different Audiences:** Consider the target audience when disseminating study findings. Tailor the communication approach to suit the needs and preferences of policymakers, healthcare professionals, patients, or the general public. Utilize plain language summaries, infographics, or presentations to engage diverse audiences.

Ensuring Reproducibility

To ensure the reproducibility of research findings, researchers should provide sufficient details regarding the study design, data collection, and analysis procedures. Consider the following practices to enhance reproducibility:

- **Provide Code and Analysis Scripts:** Include the code, syntax, or analysis scripts used for data cleaning, data analysis, and statistical modeling. Sharing the code allows other researchers to replicate the analysis and validate the findings.
- **Document Data Management:** Describe the data management procedures, including data cleaning, coding, and transformation. Document the steps taken to handle missing data, outliers, or other data quality issues.
- **Share Data and Supplementary Materials:** Consider sharing the de-identified dataset and supplementary materials, such as survey instruments, interview guides, or coding frameworks. This enables others to explore the data, conduct additional analyses, and validate the results.
- **Pre-Register Studies:** Consider pre-registering the study design, analysis plan, and outcome measures in a public registry before data collection. This practice enhances transparency and reduces the likelihood of selective reporting or post hoc analysis.

Conclusion

In this section, we explored the process of interpreting and reporting study findings in clinical research. We discussed the importance of interpreting results in the context of research objectives and existing literature, as well as effectively communicating findings through clear and concise reporting. Emphasis was placed on transparent reporting, addressing study limitations, and ensuring the reproducibility of research outcomes. By following these practices, researchers can contribute to the advancement of evidence-based practice and enhance the impact of their research within the clinical research community and beyond.

By the end of Chapter 7, readers will have a comprehensive understanding of data analysis and interpretation in clinical research. They will gain proficiency in using descriptive and inferential statistics, qualitative data analysis techniques, and exploratory data analysis to analyze and interpret study data. This knowledge will empower readers to confidently analyze their own data, draw meaningful conclusions, and effectively communicate research findings to the scientific community and other stakeholders.

8

ETHICAL CONSIDERATIONS IN CLINICAL RESEARCH

CHAPTER VIII

ETHICAL CONSIDERATIONS IN CLINICAL RESEARCH

Clinical research is a vital component of advancing medical knowledge and improving patient care. It involves the systematic investigation of new treatments, interventions, and diagnostic tools to determine their safety, efficacy, and overall impact on health outcomes. Given the involvement of human participants, clinical research is bound by strict ethical considerations to ensure the protection of participants' rights, well-being, and dignity.

SECTION 1: INTRODUCTION TO RESEARCH ETHICS

Introduction to Research Ethics

In this section, we will continue our discussion on research ethics in clinical research. We will discuss the importance of ethical considerations in ensuring the protection of research participants, maintaining scientific integrity, and upholding public trust. Readers will gain an understanding of the historical context and ethical principles that guide clinical research, including respect for autonomy, beneficence, non-maleficence, and justice. We will explore the role of ethical review boards, institutional review boards (IRBs), and research ethics committees (RECs) in overseeing and approving research protocols.

Importance of Research Ethical Considerations

At the heart of clinical research lies a commitment to advancing medical knowledge and improving patient outcomes. This noble pursuit, however, is bound by an ethical imperative to protect those who contribute to it—the participants. The introduction of this chapter illuminates the ethical principles that underpin research involving human participants, principles that ensure the integrity of research, and the welfare of participants. These principles include respect for autonomy, beneficence, non-maleficence, and justice, each serving as a compass guiding researchers through the ethical landscape of clinical studies. Ethical considerations are paramount in clinical research for several reasons:

- **Protection of Research Participants:** Ethical guidelines and regulations are designed to safeguard the rights, well-being, and safety of research participants. This includes obtaining informed consent, ensuring privacy and confidentiality, minimizing risks, and addressing any potential conflicts of interest.

- **Scientific Integrity:** Ethical research practices contribute to the scientific integrity of the research process. Adhering to ethical principles ensures that data is collected and analyzed in a rigorous and unbiased manner, leading to reliable and valid research findings.

- **Public Trust and Confidence:** Ethical research practices foster public trust and confidence in the scientific community and research enterprise. Transparency, accountability, and ethical conduct are essential for maintaining the credibility of clinical research and ensuring that research findings are used to benefit society.

Ethical Principles in Clinical Research

Ethical principles serve as a guiding framework for the conduct of research involving human participants. Four key ethical principles that underpin clinical research are respect for autonomy, beneficence, non-maleficence, and justice. Let's explore each of these principles in detail and understand how they inform the ethical conduct of research.

Respect for Autonomy

Respect for autonomy recognizes the individual's right to self-determination, freedom of choice, and respect for their decision-making capacity. It acknowledges that individuals have the right to make informed decisions about their participation in research without coercion or undue influence.

In the context of research, respect for autonomy involves:

- **Informed Consent:** Researchers must provide participants with comprehensive information about the study, including its purpose, procedures, risks, benefits, and alternatives. Participants should have the capacity to understand the information and voluntarily provide their informed consent to participate or withdraw from the study at any time.

- **Privacy and Confidentiality:** Researchers must protect the privacy and confidentiality of participants' personal information. They should ensure that participant data is securely stored, anonymized when appropriate, and used solely for research purposes.

- **Respect for Vulnerable Populations:** Special considerations are necessary when involving vulnerable populations, such as children, individuals with cognitive impairments, or those with limited decision-making capacity. Additional safeguards should be in place to protect their rights and well-being, including obtaining proxy consent from legally authorized representatives.

Beneficence and Non-maleficence

Beneficence entails acting in the best interests of participants, promoting their well-being, and maximizing benefits while minimizing risks. It involves an obligation to do good and ensure that the potential benefits of the research outweigh the potential harms.

Non-maleficence, or the principle of "do no harm," underscores the importance of preventing harm to research participants. Researchers should take all necessary precautions to minimize risks and ensure the safety and well-being of participants. This involves conducting a thorough risk-benefit assessment, implementing appropriate safety measures, and monitoring the study closely to detect and mitigate any potential harm promptly.

In the context of research, beneficence and non-maleficence involves:

- **Risk-Benefit Assessment:** Researchers should carefully assess the potential risks and benefits of the study to ensure that the anticipated benefits justify the potential harm or discomfort to participants. They should design studies with an aim to minimize risks and maximize potential benefits.

- **Monitoring Participant Safety:** Researchers have an ongoing responsibility to monitor participant safety and well-being throughout the study. Appropriate measures should be in place to identify and address any adverse events or unanticipated risks promptly.

- **Data Integrity and Quality:** Beneficence also includes conducting research with scientific rigor and integrity to generate reliable and valid results. Researchers should adhere to robust study design, appropriate data collection methods, and rigorous analysis to ensure the accuracy and credibility of findings.

Justice

The principle of justice emphasizes fairness in the distribution of benefits and burdens of research. It requires that research participants be selected in an equitable manner and that the benefits derived from research be shared fairly.

In the context of research, justice involves:

- **Selection of Participants:** Researchers should ensure that the selection of participants is fair and does not exploit or discriminate against any particular population. Inclusion criteria should be based on scientific justification and relevant considerations rather than discriminatory factors.

- **Access to Benefits:** Researchers should strive to ensure that the benefits resulting from research are accessible to all relevant populations, including those who participated in the study. This includes equitable access to any successful interventions or treatments that may emerge from the research.

- **Fair Allocation of Resources:** Justice also encompasses the fair allocation of resources for research. Researchers should consider the allocation of resources in a way that minimizes disparities and ensures equitable access to research opportunities across different populations.

- **Community Engagement:** Justice extends beyond individual participants to the broader community. Researchers should engage with the community and relevant stakeholders to understand their needs, concerns, and priorities. This engagement fosters collaboration, transparency, and accountability in the research process.

Central to these ethical considerations is the protection of participants' rights. Researchers are entrusted with the duty to safeguard the privacy and confidentiality of participant information, a task that requires diligent care and respect for the trust participants place in the research process. Ethical research practices also demand that the risks participants face are minimized and justified by the potential benefits of the research, not only to the participants themselves but also to society at large.

The introduction of this chapter sets the stage for a deeper exploration of how these ethical principles are applied in practice, guiding the conduct of clinical research. It highlights the mechanisms, challenges, and considerations that researchers face in their ethical obligations to participants. Through this exploration, readers will gain an appreciation for the complexity of ethical decision-making in clinical research and the paramount importance of upholding these ethical standards to ensure the integrity and societal value of scientific inquiry.

Historical Context of Research Ethics

Research ethics has evolved over time in response to past unethical research practices. Historical incidents, such as the Tuskegee Syphilis Study and the Nuremberg Trials, highlighted the need for ethical guidelines and regulations to protect research participants. These incidents led to the development of international ethical guidelines, codes of conduct, and regulatory frameworks that govern clinical research today.

The Tuskegee Syphilis Study

The Tuskegee Syphilis Study is a notorious episode in the history of clinical research that took place in the United States from 1932 to 1972. It involved a long-term observational study conducted by the U.S. Public Health Service (PHS) in collaboration with the Tuskegee Institute (now known as Tuskegee University) to investigate the natural progression of untreated syphilis in African American men.

Overview

The study began in 1932 in Macon County, Alabama, a predominantly African American community with limited access to healthcare. The initial aim of the study was to understand the effects of syphilis on the human body. However, the unethical nature of the study lies in the fact that the researchers intentionally withheld treatment from the participants, even after effective treatments became widely available.

Key Points:

1. **Recruitment and Deception:** The researchers recruited 600 African American men, 399 of whom had syphilis and 201 who served as a control group. The participants were not informed about the true nature of the study, instead being told they were receiving free medical care for "bad blood." This term was commonly used to refer to a variety of conditions, including syphilis.
2. **Denial of Treatment:** Despite the discovery of effective treatments like penicillin in the 1940s, the participants were denied access to these therapies. The researchers deliberately withheld treatment to observe the natural progression of the disease, even when it became clear that participants were suffering and could have benefited from treatment.
3. **Ethical Violations:** The Tuskegee Syphilis Study violated numerous ethical principles. The participants were not provided with informed consent, denying them the right to make an informed decision about their involvement in the study. Additionally, the participants were not given the option to leave the study, and they were not informed about the potential risks and benefits.
4. **Misleading Incentives:** To ensure the participants' continued participation, they were provided with incentives such as free medical examinations, meals, and burial insurance. These incentives were manipulative and exploited the participants' socioeconomic vulnerabilities.
5. **Public Disclosure and Termination:** In 1972, a whistleblower named Peter Buxtun, a PHS employee, exposed the unethical nature of the study to the press and the public. The subsequent public outrage and scrutiny led to the termination of the study, and a federal investigation was launched.

Impact and Legacy

The Tuskegee Syphilis Study had a significant impact on medical ethics, human subjects' protection, and public trust in medical research. The study exposed the need for ethical guidelines and regulations to safeguard the rights and welfare of research participants. In response to this scandal, the National Research Act was passed in 1974, leading to the establishment of institutional review boards (IRBs) and the development of ethical guidelines for human subjects' protection, such as the Belmont Report.

The legacy of the Tuskegee Syphilis Study includes increased awareness of ethical considerations in research, reforms in research practices, and a commitment to ensuring informed consent, voluntary participation, and equitable treatment of research participants. It serves as a reminder of the importance of upholding ethical principles and maintaining the trust of individuals who participate in research.

It is crucial to acknowledge the profound injustice suffered by the participants of the Tuskegee Syphilis Study and to learn from this dark chapter in the history of clinical research, ensuring that such unethical practices are never repeated.

The Nuremberg Trials

The Nuremberg Trials were a series of military tribunals held after World War II to prosecute high-ranking Nazi officials and collaborators for war crimes, crimes against humanity, and other atrocities committed during the Holocaust. The trials took place in Nuremberg, Germany, from 1945 to 1949 and set a significant precedent in international law, establishing the principles of individual accountability and informed consent for medical experimentation.

Overview

The Nuremberg Trials were a response to the heinous crimes committed by the Nazi regime, including genocide, human experimentation, and war crimes. The trials were organized by the Allied powers, represented by judges from the United States, Great Britain, France, and the Soviet Union. The main trial, known as the International Military Tribunal (IMT), focused on prosecuting 22 high-ranking Nazi officials.

Key Points:

1. **Crimes Against Humanity:** The Nuremberg Trials aimed to hold individuals accountable for their involvement in crimes against humanity. The defendants faced charges of genocide, persecution based on race or religion, war crimes, and crimes against peace. These crimes included the systematic murder of millions of Jews, Romani people, disabled individuals, and other targeted groups, as well as

forced labor, medical experimentation, and other forms of human rights violations.

2. **Medical Experiments and Informed Consent:** The Nuremberg Trials shed light on the horrific medical experiments conducted by Nazi doctors on concentration camp prisoners. These experiments violated fundamental ethical principles and included activities such as infecting prisoners with diseases, conducting freezing and hypothermia experiments, and testing the efficacy of various drugs and treatments without consent.
3. **Nuremberg Code:** In response to the unethical medical experiments, the Nuremberg Code was developed. The code laid out ten principles that formed the basis for ethical research involving human subjects. The principles emphasized voluntary informed consent, the need for scientific justification, minimizing risks, and conducting research in a way that benefits society. The Nuremberg Code has since been widely influential in shaping ethical guidelines for human subjects' protection.
4. **Legal Precedent:** The Nuremberg Trials established the principle that individuals could be held personally responsible for their actions, even if they were following orders from a higher authority. This concept, known as the "Nuremberg defense" or "superior orders defense," rejects the idea that individuals can escape accountability by claiming they were simply following orders.
5. **Legacy and Impact:** The Nuremberg Trials laid the foundation for modern international criminal law and had a profound impact on the development of ethical standards in medical research. The trials played a crucial role in shaping the Universal Declaration of Human Rights and the establishment of the International Criminal Court (ICC). They also served as a reminder of the importance of upholding ethical principles, respecting

human dignity, and protecting the rights of individuals in research.

The Nuremberg Trials represented a critical moment in history, where justice was sought for the atrocities committed during World War II. They highlighted the need for accountability and established important legal and ethical precedents. The trials not only condemned the individuals responsible for heinous crimes but also led to the formulation of the Nuremberg Code, a landmark document that continues to shape ethical guidelines for research involving human subjects. The legacy of the Nuremberg Trials serves as a constant reminder of the importance of upholding human rights, respecting individual autonomy, and ensuring that medical research is conducted ethically and with the informed consent of participants.

Role of Ethical Review Boards

Ethical review boards, such as Institutional Review Boards (IRBs) or Research Ethics Committees (RECs), play a critical role in overseeing and approving research protocols. Their primary responsibility is to ensure that research involving human participants adheres to ethical principles and meets regulatory requirements. IRBs/RECs evaluate research proposals, review study protocols, assess risks and benefits, and provide guidance to researchers. Their oversight helps protect the rights, safety, and well-being of research participants.

Conclusion

In this section, we provided a comprehensive and thorough introduction to research ethics in clinical research. We discussed the importance of ethical considerations in protecting research participants, maintaining scientific integrity, and upholding public trust. We explored ethical principles, including respect for autonomy, beneficence, non-maleficence, and justice, which guide ethical decision-making in clinical research. Additionally, we highlighted the historical context of research ethics and the pivotal role of ethical review boards, such as IRBs/RECs, in overseeing and approving research protocols. Understanding and adhering to research ethics is vital for conducting ethical and responsible clinical research.

SECTION 2: INFORMED CONSENT AND PARTICIPANT AUTONOMY: A CORNERSTONE OF ETHICAL RESEARCH

Informed Consent and Participant Autonomy: A Cornerstone of Ethical Research

This section will focus on informed consent as a cornerstone of ethical research. We will delve into the principles of informed consent, including the disclosure of information, comprehension, voluntariness, and capacity to provide consent. Readers will gain an understanding of the importance of respecting participant autonomy, providing sufficient information, and ensuring the voluntary nature of participation. We will discuss special considerations in obtaining informed consent from vulnerable populations, such as children, individuals with cognitive impairments, and individuals from culturally diverse backgrounds.

Informed Consent

Informed consent is a fundamental ethical principle and legal requirement in clinical research. It serves as a cornerstone of ethical research by upholding the principles of respect for autonomy, ensuring participant welfare, and promoting transparency and trust between researchers and participants. In this section, we will provide an extremely comprehensive understanding of informed consent and participant autonomy, covering the principles of informed consent, the components of a valid consent process, and special considerations for obtaining consent from vulnerable populations.

Principles of Informed Consent

- **Disclosure of Information:** The principle of informed consent requires researchers to provide participants with comprehensive and accurate information about the research study. This includes the purpose, procedures, potential risks and benefits, alternative options, confidentiality measures, and any compensation or incentives involved. Participants should also be informed of their right to withdraw from the study at any time.

- **Comprehension:** Informed consent necessitates that participants understand the information provided to them. Researchers should use clear and plain language, avoid technical jargon, and employ effective communication strategies to ensure participants comprehend the nature and implications of their participation. Assessing participants' understanding through comprehension checks or the opportunity to ask questions is crucial.

- **Voluntariness:** Participants must provide consent voluntarily, without coercion or undue influence. They should have the freedom to make an informed decision based on their own values, beliefs, and preferences. Researchers must avoid any form of manipulation or pressure that may compromise the voluntary nature of participants' consent.

- **Capacity to Provide Consent:** Informed consent requires that participants have the capacity to understand the information provided and make a decision regarding their participation. Assessing participants' decision-making capacity, particularly in cases involving vulnerable populations or individuals with cognitive impairments, is important to ensure that they have the ability to comprehend and make informed choices.

Special Considerations in Obtaining Informed Consent

- **Children:** When involving children in research, additional safeguards are necessary to protect their rights and well-being. Researchers must obtain informed consent from parents or legal guardians while also considering the child's assent, which involves providing age-appropriate information and involving children in the decision-making process to the extent they are capable.

- **Individuals with Cognitive Impairments:** Obtaining informed consent from individuals with cognitive impairments requires careful consideration. Researchers should adopt appropriate communication strategies, use simplified language, and involve legally authorized representatives or designated decision-makers who can act in the best interests of the individual.

- **Culturally Diverse Backgrounds:** Cultural diversity brings unique perspectives on autonomy and decision-making. Researchers should respect and accommodate cultural norms and practices, provide information in culturally sensitive ways, and consider involving community leaders or interpreters to ensure that participants from diverse backgrounds fully understand and freely provide consent.

Conclusion

Informed consent is a vital ethical principle in clinical research, upholding participant autonomy, promoting transparency, and protecting participant rights and welfare. Researchers must adhere to the principles of disclosure, comprehension, voluntariness, and capacity to provide consent when obtaining informed consent from research participants. Special considerations must be taken into account when obtaining consent from vulnerable populations, such as children, individuals with cognitive impairments, and individuals from culturally diverse backgrounds. By respecting participant autonomy and ensuring a thorough informed consent process, researchers can uphold ethical standards and foster trust in clinical research.

SECTION 3: PRIVACY, CONFIDENTIALITY, AND DATA PROTECTION: SAFEGUARDING PARTICIPANT INFORMATION

Privacy, Confidentiality, and Data Protection: Safeguarding Participant Information

In this section, we will explore the ethical considerations surrounding privacy, confidentiality, and data protection in clinical research. We will discuss the importance of safeguarding participant privacy and confidentiality, including the secure storage and transmission of sensitive data. Readers will gain an understanding of the regulatory frameworks governing data protection, such as HIPAA (Health Insurance Portability and Accountability Act) and GDPR (General Data Protection Regulation). We will explore strategies for minimizing the risk of data breaches, ensuring data anonymization, and obtaining necessary approvals for data sharing.

Privacy, Confidentiality, and Data Protections

Privacy, confidentiality, and data protection are critical ethical considerations in clinical research. These principles aim to protect participants' personal information, maintain their confidentiality, and ensure the secure handling of sensitive data. In this section, we will delve into the importance of safeguarding participant privacy and confidentiality, the regulatory frameworks that govern data protection, and strategies for minimizing the risk of data breaches and ensuring compliance with data sharing protocols.

Importance of Safeguarding Privacy and Confidentiality

- **Participant Trust:** Protecting participant privacy and confidentiality is crucial for maintaining trust in the research process. Participants are more likely to engage in research if they have confidence that their personal information will be handled securely and kept confidential.

- **Ethical Responsibility:** Researchers have an ethical responsibility to respect participants' privacy rights and protect the confidentiality of their data. This includes ensuring that only authorized personnel have access to participant information and that data are used solely for research purposes.

- **Data Integrity:** Safeguarding privacy and confidentiality contributes to data integrity by reducing the risk of bias or distortion resulting from breaches in participant confidentiality. By maintaining data security, researchers can enhance the reliability and validity of study findings.

Regulatory Frameworks

- **HIPAA (Health Insurance Portability and Accountability Act):** HIPAA sets standards for the protection of individuals' medical records and other identifiable health information. It requires researchers to obtain informed consent, establish secure data storage and transmission processes, and implement safeguards to protect participant privacy.

- **GDPR (General Data Protection Regulation):** GDPR is a European Union regulation that governs the protection of personal data. It applies to research involving EU residents and emphasizes principles of data minimization, consent, transparency, and the rights of individuals to access and control their personal data.

Minimizing Data Breach Risks and Ensuring Data Anonymization

- **Secure Data Storage and Transmission:** Researchers must employ secure methods for storing and transmitting participant data. This includes using encrypted systems, secure servers, and protected networks to minimize the risk of unauthorized access or breaches.

- **Data Anonymization:** Anonymizing participant data is a key strategy for protecting privacy and confidentiality. By removing or de-identifying identifiable information, researchers can minimize the risk of re-identification and maintain participant anonymity.

- **Data Sharing Approvals:** If researchers intend to share data with other researchers or organizations, they should obtain necessary approvals, such as data use agreements or data sharing agreements. These agreements outline the terms and conditions for data sharing, including privacy protections and limitations on data usage.

Conclusion

Privacy, confidentiality, and data protection are integral to ethical clinical research. By safeguarding participant privacy and confidentiality, researchers can foster trust, maintain data integrity, and comply with regulatory frameworks such as HIPAA and GDPR. Minimizing data breach risks, ensuring data anonymization, and obtaining appropriate approvals for data sharing are essential steps in protecting participant information. Adhering to these principles and implementing robust data protection measures contribute to the responsible and ethical conduct of clinical research.

SECTION 4: RISK-BENEFIT ASSESSMENT AND PARTICIPANT SAFETY: ETHICAL CONSIDERATIONS IN CLINICAL RESEARCH

Risk-Benefit Assessment and Participant Safety: Ethical Considerations in Clinical Research

This section will focus on the ethical considerations related to risk-benefit assessment and participant safety in clinical research. We will discuss the importance of conducting a thorough risk-benefit analysis during study design and the need to minimize potential risks to participants. Readers will gain insights into the ethical considerations involved in determining the acceptable level of risk, ensuring participant safety through appropriate monitoring and follow-up procedures, and addressing adverse events or complications. We will emphasize the need for ongoing risk assessment and the ethical responsibility of researchers to prioritize participant well-being.

Risk-Benefit Assessment

In clinical research, it is essential to conduct a thorough risk-benefit assessment and prioritize participant safety. This section focuses on the ethical considerations involved in assessing and managing risks to participants, ensuring their safety throughout the research process. By understanding the importance of risk-benefit analysis, readers will gain insights into the ethical responsibilities of researchers in minimizing potential risks, implementing safety measures, and addressing adverse events or complications.

Importance of Risk-Benefit Assessment

- **Ethical Balancing Act:** A comprehensive risk-benefit assessment is crucial in determining whether the potential benefits of a research study outweigh the risks to participants. This assessment considers the scientific value, potential impact, and relevance of the research, while also taking into account the potential harms and risks associated with participation.

- **Informed Decision-Making:** Conducting a risk-benefit analysis allows participants to make informed decisions about their involvement in the research. Providing clear and accurate information about the potential risks and benefits empowers participants to weigh their options and make choices based on their own values and preferences.

- **Scientific Integrity:** A robust risk-benefit assessment contributes to the scientific integrity of the study. By thoroughly considering potential risks and benefits, researchers can design studies that generate reliable and meaningful results while prioritizing participant safety.

Minimizing Potential Risks

- **Study Design:** Researchers should strive to minimize potential risks by carefully designing study protocols and procedures. This includes incorporating safeguards, such as eligibility criteria, monitoring mechanisms, and appropriate controls, to mitigate risks to participants.

- **Informed Consent Process:** The informed consent process plays a crucial role in ensuring participant safety. Researchers must clearly communicate the potential risks involved in the study, along with the measures in place to mitigate those risks. Participants should have a comprehensive understanding of the potential harms they may encounter.

- **Ongoing Risk Assessment:** Risk assessment should be an ongoing process throughout the research study. Researchers must continually evaluate and monitor potential risks to participants, adjusting protocols and procedures as necessary to ensure participant safety.

Participant Safety and Adverse Event Management

- **Safety Monitoring:** Researchers have an ethical responsibility to implement safety monitoring measures to identify and address adverse events or complications promptly. This may involve regular participant assessments, medical supervision, or data monitoring committees to oversee participant safety.

- **Reporting and Management of Adverse Events:** Researchers should establish clear procedures for reporting and managing adverse events or complications that may arise during the study. This includes providing appropriate medical care, documenting incidents, and reporting them to the relevant regulatory bodies and ethics committees.

- **Follow-Up and Aftercare:** Participant safety extends beyond the study period. Researchers should ensure appropriate follow-up and aftercare for participants, particularly in cases where long-term consequences or ongoing monitoring may be required.

Conclusion

Risk-benefit assessment and participant safety are paramount in ethical clinical research. By conducting a comprehensive risk-benefit analysis, researchers can balance the potential benefits of the research with the risks to participants. Minimizing potential risks, ensuring informed consent, and implementing safety measures contribute to participant safety throughout the research process. Ongoing risk assessment, prompt management of adverse events, and appropriate follow-up care are crucial components of ethical clinical research. By prioritizing participant well-being, researchers uphold the ethical principles of beneficence and non-maleficence, enhancing the credibility and integrity of their research endeavors.

SECTION 5: ETHICAL ISSUES IN DATA COLLECTION AND ANALYSIS: ENSURING RESEARCH INTEGRITY

Ethical Issues in Data Collection and Analysis: Ensuring Research Integrity

In this section, we will explore ethical issues that may arise during data collection and analysis in clinical research. We will discuss the importance of ensuring data integrity, minimizing bias, and maintaining the confidentiality of research data. Readers will gain an understanding of the ethical considerations involved in data sharing, secondary data analysis, and the use of data for commercial purposes. We will address issues related to authorship, conflicts of interest, and responsible data management practices to uphold research integrity

Ethics and Data

In the field of clinical research, ethical considerations extend beyond participant interactions and also encompass the collection and analysis of data. This section focuses on ethical issues that may arise during data collection and analysis, emphasizing the importance of upholding research integrity, minimizing bias, and maintaining data confidentiality. Readers will gain insights into the ethical considerations involved in data sharing, secondary data analysis, and the responsible use of data for commercial purposes. Additionally, we will address issues related to authorship, conflicts of interest, and responsible data management practices to ensure ethical conduct throughout the research process.

Data Integrity and Minimizing Bias

- **Data Falsification and Fabrication:** Researchers have an ethical obligation to collect and report data accurately and truthfully. Falsification or fabrication of data undermines scientific integrity and can lead to erroneous conclusions. Researchers should adhere to rigorous data collection protocols and maintain transparent and accurate records.

- **Selective Reporting and Publication Bias:** Researchers should avoid selectively reporting data that supports their hypotheses while disregarding contradictory findings. Selective reporting can introduce bias and mislead the scientific community and the public. Ethical data analysis involves presenting findings comprehensively and objectively, regardless of whether they support or contradict the initial research hypothesis.

- **Conflict of Interest:** Researchers should disclose any potential conflicts of interest that may influence the collection or analysis of data. Financial or professional conflicts of interest can compromise objectivity and lead to biased interpretation or reporting of results. Transparency and disclosure are essential to maintain research integrity.

Confidentiality and Data Privacy

- **Data Confidentiality:** Researchers must ensure the confidentiality of research data to protect participant privacy and maintain trust. Identifiable information should be securely stored, transmitted, and accessed only by authorized personnel. Researchers should implement appropriate measures to protect data confidentiality, including data encryption, access controls, and secure storage facilities.

- **Informed Consent for Data Sharing:** When data sharing is planned, researchers should seek informed consent from participants, clearly explaining how their data will be used, shared, and anonymized. Participants should have the option to provide consent specifically for data sharing, and their confidentiality should be safeguarded throughout the process.

- **Secondary Data Analysis:** When utilizing secondary data sources, such as existing databases or publicly available datasets, researchers should respect the original data contributors' consent and privacy protections. Ethical considerations include using the data for the intended purpose, adhering to data usage agreements, and protecting the confidentiality of individuals represented in the data.

Responsible Data Management and Commercialization

- **Authorship and Attribution:** Researchers should adhere to established guidelines and criteria for authorship, ensuring that individuals who have made significant contributions to the research are appropriately acknowledged as authors. Authorship should not be granted solely based on professional status or affiliation, and all authors should take responsibility for the integrity of the data and findings.

- **Commercial Use of Data:** Researchers should consider the ethical implications of using research data for commercial purposes. The potential conflicts between profit-driven interests and the responsible use of data for scientific advancement should be carefully evaluated. Transparency and adherence to ethical standards are necessary when engaging in collaborations or partnerships involving the commercial use of research data.

- **Responsible Data Management:** Researchers should adopt responsible data management practices to ensure the integrity, security, and accessibility of research data. This includes developing data management plans, implementing data backup and archiving strategies, and adhering to relevant data protection regulations. Responsible data management promotes transparency, reproducibility, and accountability.

Conclusion

Ethical issues in data collection and analysis are critical to maintaining research integrity in clinical research. Researchers have an ethical obligation to ensure data integrity, minimize bias, and maintain the confidentiality of research data. Responsible data management, appropriate data sharing practices, and transparency in reporting and authorship are essential components of ethical conduct. By addressing these ethical considerations, researchers uphold the principles of beneficence, respect for autonomy, and research integrity, fostering trust among participants, fellow researchers, and the wider scientific community.

SECTION 6: ETHICAL ISSUES IN PARTICIPANT RECRUITMENT AND RETENTION: ENSURING FAIRNESS AND ENGAGEMENT

Ethical Issues in Participant Recruitment and Retention: Ensuring Fairness and Engagement

This section will focus on the ethical considerations surrounding participant recruitment and retention in clinical research. We will discuss strategies for ensuring fairness, equity, and transparency in participant recruitment, including the avoidance of undue inducement and coercion. Readers will gain insights into the ethical challenges involved in participant retention, such as maintaining engagement and addressing potential power imbalances. We will explore the importance of cultural sensitivity, informed decision-making, and ongoing communication with participants to uphold ethical standards throughout the research process.

Ethics & Participant Recruitment

Participant recruitment and retention are crucial aspects of clinical research, and ethical considerations play a significant role in these processes. This section focuses on the ethical issues surrounding participant recruitment and retention, emphasizing the importance of fairness, equity, transparency, and participant engagement. Readers will gain insights into strategies for ethical participant recruitment, the avoidance of undue inducement and coercion, and the challenges involved in participant retention. Cultural sensitivity, informed decision-making, and ongoing communication with participants will be highlighted as essential elements in upholding ethical standards throughout the research process.

Participant Recruitment

- **Fairness and Equity:** Researchers have an ethical responsibility to ensure that participant recruitment processes are fair and equitable. This includes considering factors such as accessibility, diversity, and representation of different populations in research studies. Efforts should be made to minimize bias in recruitment and to promote inclusivity.

- **Transparency and Informed Consent:** Transparent communication about the purpose, risks, benefits, and procedures of the study is essential for informed decision-making by potential participants. Researchers should provide clear and understandable information, ensuring that participants have a comprehensive understanding of their involvement and can make autonomous decisions to participate.

- **Avoidance of Undue Inducement and Coercion:** Researchers must avoid offering excessive incentives or inducements that could unduly influence participants' decision to enroll in a study. It is crucial to ensure that participation is voluntary and not driven solely by financial gain or other external factors. Participants should feel empowered to decline participation without facing negative consequences.

Participant Retention

- **Participant Engagement and Communication:** Ethical participant retention involves maintaining ongoing communication and engagement with participants throughout the study. Researchers should provide regular updates, address participant concerns, and maintain open channels of communication to ensure participants' continued involvement and understanding of their role in the research.

- **Power Imbalances and Vulnerable Populations:** Researchers should be mindful of power imbalances that may exist between researchers and participants, especially when working with vulnerable populations. Ethical considerations include ensuring that participants have a clear understanding of their rights, maintaining a supportive and respectful environment, and addressing any potential exploitation or coercion.

- **Cultural Sensitivity and Respect:** Cultural sensitivity is crucial in participant retention, particularly when working with diverse populations. Researchers should respect participants' cultural values, norms, and beliefs, ensuring that their perspectives are acknowledged and incorporated into the research process. This includes adapting recruitment and retention strategies to be culturally appropriate and addressing potential barriers to participation.

Conclusion

Ethical participant recruitment and retention are essential for maintaining fairness, transparency, and participant autonomy in clinical research. Researchers must strive for fairness and equity in recruitment, ensuring transparency and informed decision-making. Additionally, engaging in ongoing communication, respecting cultural diversity, and addressing power imbalances contribute to ethical participant retention. By upholding these principles, researchers demonstrate their commitment to ethical conduct, participant welfare, and the overall integrity of the research process.

SECTION 7: ETHICAL RESPONSIBILITIES OF RESEARCHERS AND PUBLICATION ETHICS

Ethical Responsibilities of Researchers and Publication Ethics

In this final section, we will discuss the ethical responsibilities of researchers and publication ethics in clinical research. We will explore the principles of research integrity, including honesty, transparency, and responsible conduct of research. Readers will gain an understanding of the ethical considerations involved in authorship, plagiarism, conflicts of interest, and the responsible reporting of study findings. We will emphasize the importance of adhering to ethical guidelines, obtaining necessary approvals, and promoting scientific integrity in the dissemination of research results.

Ethics Surrounding Researchers and Publication

In this final section, we will discuss the ethical responsibilities of researchers and publication ethics in clinical research. Research integrity, honesty, transparency, and responsible conduct of research are fundamental principles that guide ethical behavior in the scientific community. Readers will gain insights into the ethical considerations related to authorship, plagiarism, conflicts of interest, and the responsible reporting of study findings. Adherence to ethical guidelines, obtaining necessary approvals, and promoting scientific integrity in the dissemination of research results will be emphasized.

Research Integrity and Responsible Conduct

- **Honesty and Transparency:** Researchers have an ethical responsibility to conduct their studies with honesty and transparency. This includes accurately reporting methods, results, and interpretations, as well as disclosing limitations or potential biases in the research process. Researchers should avoid fabricating or falsifying data, misrepresenting findings, or engaging in any form of scientific misconduct.

- **Responsible Data Management:** Ethical research involves responsible data management practices. Researchers should ensure the accuracy, integrity, and confidentiality of research data. This includes appropriately storing and securing data, documenting data collection and analysis procedures, and maintaining records for future reference and reproducibility.

- **Adherence to Ethical Guidelines:** Researchers should familiarize themselves with and adhere to ethical guidelines and regulations relevant to their research, such as those set forth by institutional review boards (IRBs) or research ethics committees (RECs). Compliance with these guidelines ensures the protection of research participants, upholds ethical standards, and maintains public trust in the research process.

Authorship and Publication Ethics

- **Authorship Criteria:** Authorship should be based on substantial contributions to the research study. Researchers should follow established authorship criteria, which typically include significant contributions to the study design, data collection, analysis, and interpretation, as well as critical involvement in drafting and revising the manuscript. All authors should have the opportunity to review and approve the final version of the manuscript.

- **Plagiarism and Attribution:** Researchers must uphold intellectual honesty and give proper credit to the works of others. Plagiarism, the uncredited use of another person's ideas, data, or work, is a serious ethical violation. Proper citation and attribution of sources are essential to acknowledge the contributions of others and maintain academic integrity.

- **Conflicts of Interest:** Researchers should disclose any conflicts of interest that could potentially bias their research or influence their objectivity. Conflicts of interest may arise from financial relationships, personal connections, or affiliations that could interfere with the unbiased reporting of research findings. Transparency in disclosing conflicts of interest is essential for maintaining research integrity.

- **Responsible Reporting and Dissemination:** Researchers have an ethical responsibility to report and disseminate their findings accurately and responsibly. This includes providing sufficient details for replication, describing study limitations, and avoiding selective reporting of results. Transparent and unbiased reporting contributes to the reliability and trustworthiness of scientific research.

Conclusion

Ethical responsibilities of researchers and publication ethics are essential in maintaining research integrity, promoting transparency, and upholding scientific standards in clinical research. Researchers must conduct their studies with honesty, transparency, and responsible data management. Adherence to ethical guidelines, appropriate authorship attribution, avoidance of plagiarism, disclosure of conflicts of interest, and responsible reporting of study findings are key ethical considerations. By adhering to these principles, researchers contribute to the overall credibility and trustworthiness of the scientific community, ensuring the dissemination of accurate and reliable research findings.

By the end of Chapter 8, readers will have a comprehensive understanding of the ethical considerations in clinical research. They will recognize the importance of informed consent, privacy protection, risk-benefit assessment, data integrity, participant recruitment and retention, and responsible conduct of research. This knowledge will empower readers to conduct research ethically, navigate complex ethical dilemmas, and contribute to the highest standards of integrity and participant protection in clinical research.

9

COLLABORATION AND TEAM SCIENCE IN CLINICAL RESEARCH

CHAPTER IX

COLLABORATION AND TEAM SCIENCE IN CLINICAL RESEARCH

This chapter explores the vital role of collaboration and team science in clinical research. This chapter emphasizes the importance of bringing together diverse expertise, perspectives, and resources to address complex scientific questions and improve healthcare outcomes. It covers the benefits and challenges of collaboration, strategies for building effective research teams, and fostering interdisciplinary cooperation. By understanding the principles of team science and the mechanisms for successful collaboration, researchers can enhance their ability to conduct comprehensive and impactful clinical studies. This chapter aims to equip readers with the knowledge and tools necessary to foster effective teamwork and collaboration in their clinical research endeavors.

SECTION 1: INTRODUCTION TO COLLABORATION IN CLINICAL RESEARCH

Introduction to Collaboration in Clinical Research

Collaboration is a fundamental aspect of clinical research that involves bringing together diverse expertise, perspectives, and resources to address complex scientific questions and improve healthcare outcomes. In today's multidisciplinary and rapidly evolving research landscape, collaboration is essential for tackling the complex challenges and advancing scientific knowledge in clinical research.

One of the key reasons why collaboration is crucial in clinical research is the breadth and depth of expertise that different team members bring to the table. By involving researchers from various disciplines, such as medicine, nursing, pharmacology, biostatistics, and social sciences, collaborative teams can leverage their collective knowledge and skills to gain comprehensive insights into the research problem at hand. For example, a clinical trial studying the effectiveness of a new cancer treatment may require collaboration among oncologists, molecular biologists, statisticians, and patient advocates to ensure a comprehensive approach.

Collaboration also enables the sharing of resources, both in terms of infrastructure and data. Research institutions and organizations often possess specialized equipment, databases, and clinical cohorts that can be shared among collaborators, leading to cost efficiencies and broader access to valuable resources. This sharing of resources facilitates larger-scale studies, enables more robust analyses, and accelerates the pace of research. For instance, a consortium of hospitals collaborating on a multi-center study can pool their patient data and infrastructure to increase sample size and generate more reliable findings.

Furthermore, collaboration promotes innovation by fostering the exchange of ideas and perspectives. When researchers with diverse backgrounds and experiences come together, they can generate novel insights, challenge assumptions, and explore innovative approaches to research questions. For example, a collaboration between geneticists and bioengineers might lead to the development of new genetic sequencing technologies or innovative drug delivery systems.

However, effective collaboration in clinical research also comes with its challenges. Coordinating efforts, managing diverse opinions, and aligning goals and expectations among team members can be complex. Effective communication and leadership are crucial to ensure a shared vision, establish clear roles and responsibilities, and maintain transparency throughout the research process.

In summary, collaboration plays a vital role in clinical research by leveraging diverse expertise, sharing resources, and promoting innovation. Through collaboration, researchers can address complex research questions, improve the validity and generalizability of findings, and ultimately enhance patient care. By working together as a team, the potential for interdisciplinary discoveries and advancements in clinical research is greatly amplified.

SECTION 2: BUILDING EFFECTIVE RESEARCH TEAMS

Building Effective Research Teams

This section will focus on building effective research teams. We will discuss the key elements of successful research teams, including clear communication, shared goals, and complementary expertise. Readers will gain insights into strategies for team formation, such as identifying team members with diverse skills, establishing team roles and responsibilities, and fostering a positive team culture. We will explore the importance of effective leadership, conflict resolution, and promoting inclusivity within research teams.

Building an effective research team is crucial for the success of collaborative clinical research endeavors. A well-functioning team possesses key elements that contribute to its productivity, synergy, and ability to achieve research goals. In this section, we will explore the essential components of successful research teams and provide strategies for team formation and maintenance.

Clear Communication

Effective communication is the cornerstone of any successful research team. Team members must establish open lines of communication, both within the team and with external collaborators or stakeholders. Clear and regular communication ensures that everyone is aligned with project objectives, research progress, and any challenges or updates. This can be achieved through regular team meetings, shared project management platforms, and the use of clear and concise communication channels. For example, a research team may utilize project management software to assign tasks, track progress, and facilitate seamless communication among team members.

Shared Goals and Vision

Research teams should have a shared understanding of the project goals and a common vision for the research outcomes. This shared purpose serves as a driving force that motivates team members and fosters collaboration. Clearly defined goals and objectives help align individual efforts toward a common research agenda. For instance, a team working on a clinical trial may have a shared goal of evaluating a new treatment's efficacy in improving patient outcomes, which guides their collective efforts.

Complementary Expertise

A diverse range of expertise within the team is essential for comprehensive and interdisciplinary research. Each team member should bring unique skills, knowledge, and perspectives that complement one another. By combining expertise from various disciplines, such as medicine, biostatistics, psychology, and sociology, research teams can address complex research questions from multiple angles. For example, a team investigating the impact of a new intervention for mental health might include psychiatrists, psychologists, social workers, and statisticians, each contributing their specialized knowledge.

Team Formation and Roles

Thoughtful team formation involves identifying individuals with the necessary skills and expertise to contribute to the research project. Assigning specific roles and responsibilities to team members ensures clarity in task distribution and accountability. Team leaders should consider individuals' strengths and expertise when assigning roles, aiming for a balanced allocation of responsibilities. For instance, a team may have a project manager responsible for overall coordination, a statistician handling data analysis, and a clinician providing clinical expertise.

Positive Team Culture

Fostering a positive team culture is vital for promoting collaboration and maintaining team morale. This includes cultivating an environment that encourages open dialogue, respect, and trust among team members. Recognizing and appreciating individual contributions, celebrating milestones, and promoting a sense of camaraderie can contribute to a positive team culture. For example, regular team-building activities, such as team lunches or social events, can foster a sense of unity and cohesion.

Effective Leadership and Conflict Resolution

Strong leadership is crucial for guiding and coordinating the research team. Effective leaders provide clear direction, facilitate decision-making, and encourage collaboration. They also play a critical role in conflict resolution, addressing any disagreements or issues that may arise within the team. Leaders should create an environment where conflicts are managed constructively, allowing for open dialogue and finding mutually agreeable solutions.

Promoting Inclusivity and Diversity

Research teams should strive to be inclusive and diverse, encompassing individuals from different backgrounds, cultures, and perspectives. Embracing diversity enriches the team's collective knowledge, fosters creativity, and promotes a broader understanding of research questions and outcomes. Inclusive practices ensure that all team members feel valued, respected, and have equal opportunities to contribute.

In conclusion, building an effective research team requires clear communication, shared goals, complementary expertise, effective leadership, and a positive team culture. By focusing on these key elements, research teams can maximize collaboration, productivity, and the generation of high-quality research outcomes. Examples such as multidisciplinary teams comprising clinicians, statisticians, and social scientists collaborating on a community-based health intervention or a team of researchers from different countries working together to address a global health issue can further illustrate the importance and benefits of building effective research teams.

SECTION 3: INTERDISCIPLINARY COLLABORATION

Interdisciplinary Collaboration

In this section, we will delve into the importance of interdisciplinary collaboration in clinical research. We will discuss the value of bringing together researchers from different disciplines, such as medicine, nursing, psychology, and statistics, to address complex research questions. Readers will gain an understanding of the benefits of interdisciplinary collaboration, including a broader perspective, innovative approaches, and the potential for more comprehensive and impactful research outcomes. We will explore strategies for promoting interdisciplinary collaboration, such as creating interdisciplinary research centers, fostering cross-disciplinary training, and establishing collaborative networks.

Interdisciplinary collaboration plays a pivotal role in advancing clinical research by bringing together researchers from diverse disciplines to collectively address complex research questions and challenges. In this section, we will delve into the importance of interdisciplinary collaboration in clinical research, highlight its benefits, and provide strategies for promoting and facilitating such collaborations.

Broad Perspective and Comprehensive Understanding

Interdisciplinary collaboration allows researchers from different disciplines, such as medicine, nursing, psychology, and statistics, to contribute their unique expertise and perspectives to a research project. By integrating multiple disciplinary viewpoints, researchers can develop a more comprehensive understanding of the research problem, considering various biological, psychological, social, and statistical factors. For example, in a study examining the impact of a new treatment for a specific condition, interdisciplinary collaboration can provide insights into the medical efficacy, psychological well-being of patients, and statistical analysis of outcomes, leading to a more holistic understanding.

Innovative Approaches and Solutions

The convergence of diverse disciplines fosters innovation by encouraging the exchange of ideas, methodologies, and technologies. Researchers with different backgrounds bring distinct approaches to problem-solving and can inspire novel research methodologies and interventions. For instance, collaborations between engineers and medical researchers have led to the development of innovative medical devices and technologies, revolutionizing patient care.

Enhanced Study Design and Analysis

Interdisciplinary collaboration strengthens study design and analysis by incorporating expertise from various disciplines. Researchers from different fields can contribute to the development of research protocols, ensuring that relevant variables and outcomes are considered. Moreover, interdisciplinary collaboration improves data analysis by incorporating advanced statistical techniques tailored to the research question. By leveraging the expertise of statisticians, researchers can apply sophisticated analytical methods to extract meaningful insights from complex datasets.

Translation of Research into Practice

Interdisciplinary collaboration facilitates the translation of research findings into clinical practice and policy. Collaboration with health-care professionals and policymakers ensures that research outcomes align with real-world needs and can be effectively implemented. For example, collaborations between researchers and healthcare providers can facilitate the translation of research findings into evidence-based guidelines or interventions that directly impact patient care.

Strategies for Promoting Interdisciplinary Collaboration

- **Interdisciplinary Research Centers:** Establishing interdisciplinary research centers or institutes fosters collaboration by providing a dedicated space for researchers from different disciplines to collaborate on common research goals. These centers can facilitate regular interactions, knowledge exchange, and joint funding opportunities. For example, a research center focusing on cardiovascular health might bring together experts in cardiology, epidemiology, behavioral science, and health economics.

- **Cross-Disciplinary Training:** Promoting cross-disciplinary training programs encourages researchers to acquire knowledge and skills beyond their primary discipline. These programs can include workshops, seminars, or graduate courses that expose researchers to different disciplines and facilitate collaboration across fields. For instance, a training program on translational research may bring together clinicians, basic scientists, and public health researchers to foster interdisciplinary collaboration.

- **Collaborative Networks:** Establishing collaborative networks and platforms encourages researchers to connect with colleagues from different disciplines. These networks can facilitate communication, knowledge sharing, and collaboration on research projects. Online platforms, conferences, and interdisciplinary research consortia can serve as avenues for researchers to identify potential collaborators and form interdisciplinary teams.

- **Funding Opportunities:** Funding agencies and organizations can play a crucial role in promoting interdisciplinary collaboration by providing grants and funding specifically designated for interdisciplinary research. By allocating resources to interdisciplinary projects, funding agencies incentivize researchers to form interdisciplinary teams and tackle complex research questions that require diverse expertise.

Examples of Interdisciplinary Collaboration:

The field of precision medicine brings together researchers from genetics, bioinformatics, clinical medicine, and statistics to develop personalized treatment approaches based on an individual's genetic makeup and other factors.

The integration of behavioral science and public health research aims to understand the social and psychological determinants of health and design interventions that address these factors.

Collaborations between engineers, material scientists, and biomedical researchers have led to advancements in tissue engineering, such as the development of artificial organs and bioengineered scaffolds for tissue regeneration.

Interdisciplinary collaborations between neuroscientists, computer scientists, and psychologists have resulted in the development of advanced brain imaging techniques and computational models to study cognitive processes and neurological disorders.

To summarize, interdisciplinary collaboration in clinical research offers numerous benefits, including a broader perspective, innovative approaches, and comprehensive research outcomes. By promoting interdisciplinary collaboration through various strategies and examples, researchers can harness the collective expertise of different disciplines and advance scientific knowledge for improved patient outcomes and population health.

SECTION 4: COLLABORATING WITH INDUSTRY AND EXTERNAL PARTNERS

Collaborating with Industry and External Partners

This section will focus on the importance of collaborating with industry and external partners in clinical research. We will discuss the benefits of industry collaborations, such as access to resources, expertise, and opportunities for translational research. Readers will gain insights into strategies for establishing successful collaborations with industry, including research agreements, intellectual property considerations, and effective communication between academia and industry partners. We will explore the ethical considerations, potential conflicts of interest, and best practices for maintaining scientific independence and transparency in industry collaborations.

Collaborating with industry and external partners in clinical research plays a vital role in advancing scientific knowledge, translating research findings into practical applications, and improving patient care. In this section, we will explore the benefits of industry collaborations, discuss strategies for establishing successful partnerships, and address the ethical considerations associated with these collaborations.

Benefits of Industry Collaborations

- **Access to Resources:** Industry collaborations provide researchers with access to a wide range of resources, including funding, state-of-the-art facilities, specialized equipment, and established networks. These resources can significantly enhance the research infrastructure and accelerate the progress of clinical research projects. For example, pharmaceutical companies may provide financial support for clinical trials, making it possible to conduct large-scale studies that would otherwise be financially challenging for academic institutions.

- **Expertise and Specialized Knowledge:** Collaborating with industry partners allows researchers to tap into the expertise and specialized knowledge of professionals working in the commercial sector. Industry partners bring a wealth of experience in product development, regulatory affairs, and commercialization strategies. By collaborating with industry, researchers can gain valuable insights and guidance in navigating the complexities of translating research discoveries into marketable products or interventions.

- **Translational Research Opportunities:** Industry collaborations offer opportunities for translational research, bridging the gap between basic research and clinical applications. By working closely with industry partners, researchers can explore the potential for translating their findings into practical solutions, such as new therapies, medical devices, or diagnostic tools. These collaborations facilitate the development of evidence-based interventions that can directly benefit patients and improve healthcare outcomes.

Strategies for Establishing Successful Collaborations with Industry

- **Research Agreements and Contracts:** Clear research agreements and contracts are essential to establish the terms of collaboration, including intellectual property rights, data sharing, publication policies, and financial considerations. These agreements should define the roles, responsibilities, and expectations of both academic and industry partners to ensure a mutually beneficial collaboration.

- **Effective Communication and Collaboration:** Open and effective communication between academia and industry partners is crucial for successful collaborations. Regular meetings, progress updates, and collaborative decision-making processes facilitate transparency and foster a shared understanding of project goals and milestones. It is important to establish clear lines of communication and designated points of contact to streamline collaboration efforts.

- **Intellectual Property Considerations:** Intellectual property (IP) is an important consideration in industry collaborations. Researchers and institutions should define ownership and management of IP rights through agreements to protect their research discoveries and ensure appropriate commercialization. Clear guidelines for the disclosure, protection, and licensing of IP should be established early in the collaboration.

- **Maintain Scientific Independence and Transparency:** While industry collaborations bring valuable resources and expertise, it is crucial to maintain scientific independence and transparency. Researchers must uphold scientific integrity, adhere to ethical guidelines, and ensure that research findings are reported accurately and objectively. Transparent reporting of research results, regardless of outcomes, is essential to maintain public trust and scientific credibility.

Ethical Considerations and Best Practices

- **Conflicts of Interest:** Collaboration with industry may give rise to potential conflicts of interest. Researchers must disclose any financial or personal interests that could potentially influence the design, conduct, or reporting of their research. Institutions should have clear policies and procedures in place to manage and mitigate conflicts of interest, ensuring that research integrity is not compromised.

- **Protection of Human Subjects:** Industry collaborations involving human subjects require ethical considerations and adherence to relevant regulations and guidelines. Researchers must ensure the protection of participants' rights, privacy, and well-being throughout the research process. Ethical review boards or institutional review boards should provide oversight and approval of research protocols involving human subjects.

- **Data Sharing and Publication:** Collaborations should establish clear guidelines for data sharing and publication. Researchers should determine the extent and timing of data sharing, taking into account any proprietary or confidential information. Timely publication of research findings in peer-reviewed journals is essential for disseminating knowledge and promoting transparency.

Examples of Industry Collaborations:

1. A pharmaceutical company collaborates with academic researchers to conduct clinical trials for a new drug. The company provides funding, expertise in drug development and regulatory affairs, while the academic researchers contribute their clinical expertise and access to patient populations.
2. An academic research institute partners with a medical device company to develop and test a novel medical device for a specific disease. The collaboration involves joint research and development efforts, sharing of technical expertise, and access to the company's manufacturing facilities.
3. A digital health startup collaborates with a university research center to evaluate the effectiveness of a mobile health app for managing chronic conditions. The collaboration involves data collection, analysis, and validation of the app's efficacy, leveraging the startup's technological expertise and the research center's clinical research capabilities.

Conclusion

Collaborating with industry and external partners brings numerous benefits to clinical research, including access to resources, expertise, and translational opportunities. By establishing successful collaborations through clear agreements, effective communication, and ethical considerations, researchers can harness the strengths of both academia and industry to drive innovation and improve patient outcomes. However, it is crucial to maintain scientific independence, transparency, and ethical integrity to ensure that research findings remain unbiased and impactful.

SECTION 5: PATIENT AND PUBLIC INVOLVEMENT IN RESEARCH

Patient and Public Involvement in Research

In this section, we will delve into the importance of involving patients and the public in clinical research. We will discuss the benefits of patient and public involvement, including increased relevance, improved research design, and enhanced dissemination and implementation of research findings. Readers will gain an understanding of strategies for engaging patients and the public in the research process, such as involving them in research design, recruitment, and data interpretation. We will explore the ethical considerations, challenges, and best practices for meaningful patient and public involvement in research.

Patient and public involvement in clinical research is essential for ensuring that research is relevant, meaningful, and effectively translated into practice. In this section, we will explore the importance of involving patients and the public in the research process, discuss the benefits of their involvement, and provide strategies for meaningful engagement.

Benefits of Patient and Public Involvement

- **Increased Relevance:** Involving patients and the public in research ensures that studies address their needs and priorities. Their lived experiences and perspectives provide valuable insights that can shape research questions, study design, and outcomes of interest. By involving those directly affected by the condition or intervention under investigation, researchers can ensure that the research is meaningful and addresses real-world concerns.

- **Improved Research Design:** Patients and the public can contribute to the development of research protocols and study materials. Their input can help refine research questions, determine appropriate study endpoints, and identify patient-centered outcomes. Involving them in the design phase can enhance the feasibility and acceptability of the research, leading to more successful recruitment and retention of participants.

- **Enhanced Dissemination and Implementation:** Engaging patients and the public throughout the research process improves the communication and dissemination of research findings. Their involvement ensures that study results are communicated in a way that is accessible, understandable, and actionable for the wider community. Patient and public perspectives can also aid in identifying effective strategies for implementing research findings into clinical practice and policy.

Strategies for Engaging Patients and the Public

- **Involvement in Research Design:** Patients and the public can be involved in setting research priorities, refining research questions, and determining study outcomes. This can be achieved through advisory panels, focus groups, or workshops where their input is sought and incorporated into the research design.

- **Recruitment and Participant Engagement:** Patients and the public can contribute to the development of participant recruitment materials, such as information sheets and consent forms. Their involvement can help ensure that these materials are clear, informative, and respectful of participants' needs. Additionally, involving patients and the public in participant engagement activities, such as study newsletters or patient advisory boards, can foster ongoing collaboration and feedback.

- **Data Interpretation and Dissemination:** Including patients and the public in data interpretation and dissemination activities can enhance the relevance and impact of research findings. Their perspectives can provide valuable insights into the meaning and implications of study results. Collaborative efforts can include co-authorship of research publications, involvement in lay summaries, and active participation in conferences or public presentations.

Ethical Considerations and Best Practices

- **Representation and Diversity:** It is important to ensure diversity and representation among the patients and public involved in research. Efforts should be made to engage individuals from different demographics, socio-economic backgrounds, and cultural groups to ensure that a range of perspectives is considered.

- **Training and Support:** Providing training and support to patients and the public involved in research is crucial for their effective engagement. This may include education on research processes, ethics, and the role they will play in the study. Researchers should ensure that participants have access to appropriate resources and support throughout their involvement.

- **Recognition and Compensation:** Recognizing the contributions of patients and the public is essential. Researchers should acknowledge their involvement in research activities and consider appropriate forms of compensation or reimbursement for their time and effort.

Examples of Patient and Public Involvement:

1. A research project on mental health services involves a patient advisory group consisting of individuals with lived experience of mental illness. The group meets regularly to provide input on research priorities, study design, and patient-centered outcomes. They actively contribute to the development of research materials and participate in the interpretation and dissemination of findings.
2. A clinical trial investigating a new treatment for a rare disease involves engaging patient advocacy groups in the

recruitment process. These groups help identify potential participants, provide support to enrolled participants, and contribute to the dissemination of trial results to the wider patient community.

3. A research institute hosts public forums where community members are invited to discuss ongoing research projects and provide input on future research directions. These forums facilitate open dialogue between researchers and the public, allowing for a shared understanding of research goals and fostering collaboration.

Conclusion

Involving patients and the public in clinical research is vital for its relevance, quality, and impact. Their involvement brings unique perspectives, enhances research design, and improves the dissemination and implementation of research findings. By adopting strategies for meaningful engagement, considering ethical considerations, and recognizing their contributions, researchers can harness the power of patient and public involvement to drive positive change in clinical research and healthcare.

SECTION 6: COLLABORATIVE RESEARCH FUNDING AND GRANTSMANSHIP

Collaborative Research Funding and Grantsmanship

This section will focus on collaborative research funding and grantsmanship. We will discuss the importance of securing collaborative research grants and funding to support interdisciplinary and collaborative research endeavors. Readers will gain insights into strategies for identifying funding opportunities, developing collaborative research proposals, and establishing a strong research team. We will explore the role of research institutions, funding agencies, and consortia in promoting collaborative research and providing resources and support for successful grant applications

Collaborative research funding and grantsmanship play a crucial role in supporting interdisciplinary and collaborative research endeavors. In this section, we will explore the importance of securing collaborative research grants and funding, discuss strategies for identifying funding opportunities and developing strong research proposals, and highlight the role of research institutions and funding agencies in promoting collaborative research.

Importance of Collaborative Research Funding

- **Resource Mobilization:** Collaborative research often requires significant resources, including funding for personnel, equipment, data collection, and analysis. Collaborative research grants provide the necessary financial support to undertake ambitious and large-scale research projects that may be beyond the capacity of individual researchers or institutions.

- **Interdisciplinary Collaboration:** Collaborative research grants facilitate interdisciplinary collaboration by bringing together researchers from diverse fields to address complex research questions. Such collaborations foster cross-pollination of ideas, expertise, and methodologies, leading to innovative approaches and breakthrough discoveries that may not be possible through individual efforts.

- **Enhanced Research Impact:** Collaborative research funded through grants has the potential for broader and more significant impact. By leveraging the strengths and resources of multiple research teams, collaborative projects can generate findings that have implications across disciplines, influence policy, and contribute to improved patient outcomes and societal well-being.

Strategies for Securing Collaborative Research Funding

- **Identifying Funding Opportunities:** Researchers should actively explore funding sources that explicitly support collaborative research initiatives. This may include government agencies, philanthropic foundations, industry partnerships, and international funding programs. Research institutions and online databases can provide information on available funding opportunities.

- **Developing Strong Research Proposals:** Effective grant proposals for collaborative research require careful planning and coordination. Researchers should clearly articulate the research objectives, the rationale for collaboration, and the expected outcomes and impact. The proposal should demonstrate the complementarity of expertise among team members and describe the collaborative processes and structures that will be in place.

- **Establishing a Strong Research Team:** Building a strong research team is essential for successful collaborative research funding. Researchers should identify team members with diverse skills and expertise that align with the research objectives. Collaborative relationships and past research collaborations can strengthen the proposal's credibility and demonstrate the capacity to work effectively as a team.

Role of Research Institutions and Funding Agencies

Research institutions and funding agencies play a crucial role in promoting collaborative research and providing support for successful grant applications. They offer various resources and initiatives to facilitate collaboration and enhance the competitiveness of research proposals. Examples include:

- **Research Consortia:** Research institutions may establish consortia or interdisciplinary centers that bring together researchers from different disciplines to foster collaboration, provide infrastructure support, and facilitate access to funding opportunities.

- **Funding Programs:** Funding agencies often have specific funding programs dedicated to collaborative research. These programs may provide grants for interdisciplinary projects, team-based research, or initiatives that address complex societal challenges. They may also offer workshops, training, and networking events to support researchers in developing collaborative proposals.

- **Research Support Services:** Research institutions provide support services to assist researchers in grant development, budgeting, and project management. These services may include grant writing workshops, access to proposal development resources, and assistance with administrative processes.

- **Partnership Development:** Research institutions and funding agencies actively promote partnerships between academia, industry, and other stakeholders. They facilitate collaborations by connecting researchers with potential industry partners, facilitating knowledge exchange, and supporting technology transfer and commercialization efforts.

Example: A research institution establishes a collaborative research center focused on sustainable energy solutions. The center brings together researchers from engineering, environmental science, economics, and policy disciplines. It offers dedicated funding opportunities for collaborative projects, hosts interdisciplinary workshops, and provides technical and administrative support to researchers involved in the center.

Conclusion

Securing collaborative research funding is vital for supporting interdisciplinary research efforts. By identifying funding opportunities, developing strong research proposals, and leveraging the resources and support provided by research institutions and funding agencies, researchers can enhance their chances of securing funding for collaborative research. Collaboration not only enriches research outcomes but also promotes innovation, cross-disciplinary learning, and societal impact.

SECTION 7: OVERCOMING CHALLENGES AND ENHANCING COLLABORATION

Overcoming Challenges and Enhancing Collaboration

In this final section, we will discuss common challenges in collaboration and strategies for enhancing collaboration in clinical research. We will explore challenges such as communication barriers, conflicting priorities, and power dynamics within research teams. Readers will gain an understanding of strategies for fostering effective collaboration, including establishing clear communication channels, promoting mutual respect and trust, and leveraging technology for virtual collaborations. We will emphasize the importance of ongoing evaluation, feedback, and continuous improvement in enhancinq collaborative research endeavors.

In this final section, we will discuss the common challenges that researchers face in collaboration and strategies for overcoming these challenges to enhance collaboration in clinical research. Collaboration is a complex process that requires effective communication, alignment of priorities, and the ability to navigate power dynamics within research teams. By addressing these challenges and implementing strategies for enhancing collaboration, researchers can foster a more productive and harmonious research environment.

Common Challenges in Collaboration

- **Communication Barriers:** Communication breakdowns, language barriers, and differences in communication styles can hinder effective collaboration. Misunderstandings and lack of clarity can lead to delays, conflicts, and inefficiencies in research projects.
- **Conflicting Priorities:** Collaborative research often involves multiple stakeholders with different priorities, interests, and timelines. Conflicting priorities can create tension and hinder progress if not properly managed.
- **Power Dynamics:** Power imbalances within research teams, such as hierarchical structures or disparities in expertise, can impact decision-making processes and hinder open and inclusive collaboration.

Strategies for Enhancing Collaboration

- **Clear Communication Channels:** Establishing clear and open lines of communication among team members is essential. Regular team meetings, clear communication protocols, and utilizing collaboration tools and technologies can facilitate effective communication and ensure that information flows smoothly among team members.
- **Mutual Respect and Trust:** Fostering an environment of mutual respect and trust is crucial for effective collaboration. Team members should value and appreciate each other's contributions, expertise, and perspectives. Encouraging open dialogue, active listening,

and valuing diverse viewpoints can strengthen collaboration and build trust among team members.

- **Collaboration Agreements:** Developing collaboration agreements or charters that outline shared goals, roles, responsibilities, and expectations can promote a common understanding among team members. These agreements can serve as a reference point to manage conflicts and ensure everyone is aligned towards the project's objectives.

- **Conflict Resolution:** Conflict is inevitable in collaborative endeavors. Establishing a process for conflict resolution, such as mediation or designated conflict resolution facilitators, can help address disagreements and ensure that conflicts are resolved in a constructive and timely manner.

- **Technology and Virtual Collaboration:** In an increasingly interconnected world, technology plays a vital role in facilitating collaboration. Utilizing virtual collaboration tools, project management software, and video conferencing platforms can enable remote collaboration, bridge geographical barriers, and enhance communication and collaboration among team members.

- **Ongoing Evaluation and Continuous Improvement:** Regularly evaluating the collaboration process and seeking feedback from team members can help identify areas for improvement. Implementing lessons learned from previous collaborations, conducting post-project reviews, and fostering a culture of continuous improvement can enhance future collaborative research endeavors.

Example: A research team conducting a multi-site clinical trial establishes a project management platform to streamline communication and document sharing among team members. They hold regular virtual meetings to discuss project progress, address any challenges, and ensure that all team members are aligned with project goals. The team also conducts periodic evaluations to assess the effectiveness of collaboration and implements adjustments based on feedback received.

Conclusion

Collaboration in clinical research can be challenging but essential for successful outcomes. By addressing common challenges such as communication barriers, conflicting priorities, and power dynamics, and implementing strategies for enhancing collaboration, researchers can create a more inclusive, efficient, and productive research environment. Emphasizing clear communication channels, mutual respect and trust, and leveraging technology can contribute to overcoming challenges and fostering effective collaboration in clinical trials. Ongoing evaluation and a culture of continuous improvement further enhance the collaborative research process.

By the end of Chapter 9, readers will have a comprehensive understanding of the importance of collaboration and team science in clinical research. They will recognize the benefits of interdisciplinary collaboration, industry partnerships, patient and public involvement, and collaborative funding. This knowledge will empower researchers to build effective research teams, establish successful collaborations, and foster a culture of collaboration and team science in their clinical research endeavors.

10

MANAGING CHALLENGES IN CLINICAL RESEARCH

CHAPTER X

MANAGING CHALLENGES IN CLINICAL RESEARCH

Chapter 10 addresses the myriad challenges inherent in conducting clinical research. This chapter provides a comprehensive exploration of the obstacles researchers face, from resource constraints and regulatory complexities to data management and interdisciplinary collaboration. By identifying and understanding these challenges, researchers can develop effective strategies to overcome them, ensuring the successful implementation and integrity of their studies. Emphasis is placed on proactive problem-solving, adaptability, and continuous learning to navigate the evolving landscape of clinical research. Through this detailed examination, readers will be equipped with the knowledge and tools necessary to manage and mitigate the difficulties encountered in the dynamic field of clinical research.

SECTION 1: INTRODUCTION TO CHALLENGES IN CLINICAL RESEARCH

Introduction to Challenges in Clinical Research

In the field of clinical research, researchers often face various challenges that can impact the successful implementation of their projects. These challenges arise due to the dynamic nature of research environments, resource constraints, regulatory complexities, and the evolving landscape of healthcare. In this section, we will explore these challenges and highlight the importance of identifying and managing them effectively.

Dynamic Research Environments

Clinical research takes place in dynamic environments where factors such as advancements in technology, changes in healthcare policies, and emerging scientific knowledge can significantly influence research processes. Researchers must stay updated with the latest developments in their respective fields to ensure that their study designs, methodologies, and data collection strategies align with the current best practices. Flexibility and adaptability are crucial in navigating the evolving landscape of clinical research.

Resource Constraints

One of the primary challenges in clinical research is resource limitations, including funding, staffing, and infrastructure. Conducting high-quality research requires adequate financial resources to support study activities, such as participant recruitment, data collection, analysis, and publication. Researchers must navigate these resource constraints by seeking funding opportunities, optimizing research team efficiency, and collaborating with other institutions or research networks to leverage shared resources.

Regulatory Complexities

Clinical research is subject to regulatory frameworks that aim to protect the rights, safety, and well-being of research participants. Researchers must navigate a complex landscape of ethical guidelines, institutional review boards (IRBs), research ethics committees (RECs), and regulatory requirements specific to their study settings. Compliance with these regulations and obtaining necessary approvals can be time-consuming and challenging. Researchers must dedicate sufficient time and effort to ensure ethical and regulatory compliance throughout their research projects.

Evolving Healthcare Landscape

The healthcare landscape is continuously evolving, with advancements in medical technology, changes in healthcare delivery models, and evolving treatment guidelines. Researchers must consider these changes when designing and implementing clinical research studies. Adapting research protocols to align with current clinical practices and addressing emerging research questions that reflect the evolving healthcare needs are essential. Collaboration with healthcare professionals and stakeholders can provide valuable insights into these evolving trends and challenges.

Conclusion

Clinical research presents various challenges that researchers must navigate to ensure the successful implementation of their projects. The dynamic nature of research environments, resource constraints, regulatory complexities, and the evolving healthcare landscape demand researchers' adaptability and effective management strategies. By recognizing and proactively addressing these challenges, researchers can optimize the research process, enhance the quality and impact of their studies, and contribute to advancements in clinical knowledge and patient care.

SECTION 2: PLANNING AND DESIGN CHALLENGES

Planning and Design Challenges

The planning and design phase of clinical research studies involves several challenges that researchers need to address effectively. In this section, we will explore key challenges related to defining research objectives, selecting appropriate study designs, estimating sample sizes, study feasibility, recruitment, and retention. By understanding and proactively addressing these challenges, researchers can enhance the quality and success of their studies.

Defining Research Objectives

One of the initial challenges in clinical research is clearly defining research objectives. Researchers must identify the research question or hypothesis they aim to address and articulate specific and measurable objectives. Ambiguous or broad research objectives can lead to difficulties in study design, data collection, and data analysis. It is essential to conduct a thorough literature review, consult with subject matter experts, and engage key stakeholders to ensure the research objectives are well-defined and aligned with the research goals.

Study Design Selection

Choosing an appropriate study design is critical for generating valid and reliable research findings. Researchers must consider factors such as the research question, available resources, ethical considerations, and feasibility when selecting a study design. Challenges may arise in deciding between observational or experimental designs, cross-sectional or longitudinal approaches, or selecting the most suitable control or comparison groups. Researchers should carefully evaluate the strengths and limitations of different study designs to ensure they align with the research objectives and can effectively address the research question.

Estimating Sample Sizes

Accurate estimation of sample sizes is essential for ensuring statistical power and detecting meaningful effects in clinical research. Insufficient sample sizes may result in underpowered studies that fail to detect significant differences or associations. Conversely, overly large sample sizes can lead to resource wastage and ethical concerns. Researchers need to carefully consider factors such as the anticipated effect size, desired statistical power, alpha level, and variability in outcomes to estimate appropriate sample sizes. Statistical power analysis should be conducted a priori to guide sample size determination.

Study Feasibility

Ensuring the feasibility of a research study is crucial for its successful implementation. Researchers may face challenges related to available resources, time constraints, and logistical considerations. It is essential to assess the feasibility of study procedures, such as participant recruitment, data collection methods, and follow-up assessments. Conducting a pilot study or feasibility assessment can help identify and address potential challenges early in the planning phase, allowing for necessary adjustments and refinements to the study design.

Recruitment and Retention

Recruiting and retaining an adequate number of participants throughout the study duration is a common challenge in clinical research. Low participant enrollment and high attrition rates can compromise the statistical power and generalizability of study findings. Researchers must develop effective recruitment strategies, such as engaging with relevant clinical networks, employing diverse recruitment methods, and addressing barriers to participation. Similarly, implementing strategies to enhance participant retention, such as regular communication, incentives, and participant-centered approaches, can help minimize attrition and improve data quality.

Stakeholder Engagement

Engaging key stakeholders, including patients, healthcare providers, community organizations, and research staff, is crucial for successful study planning and design. Challenges may arise in establishing effective collaborations, obtaining buy-in from stakeholders, and aligning research objectives with the needs and perspectives of relevant stakeholders. Researchers should involve stakeholders early in the planning phase, seek their input and feedback, and ensure their perspectives are considered throughout the study. Engaging stakeholders can enhance the relevance and applicability of the research, improve participant recruitment and retention, and facilitate knowledge translation and dissemination.

Conclusion

Planning and designing clinical research studies entail various challenges that researchers must navigate to ensure study success. By addressing challenges related to defining research objectives, selecting appropriate study designs, estimating sample sizes, study feasibility, recruitment, and retention, researchers can optimize the quality and impact of their studies. Thorough planning, piloting, stakeholder engagement, and flexibility in adapting to potential challenges contribute to the successful implementation of clinical research projects and the generation of valuable scientific knowledge.

SECTION 3: RESOURCE MANAGEMENT CHALLENGES

Resource Management Challenges

The effective management of resources is crucial for the successful implementation of clinical research projects. In this section, we will explore the challenges associated with resource management, including funding limitations, staff allocation, and infrastructure requirements. We will discuss strategies for securing funding, optimizing resource allocation, and leveraging collaborations to overcome resource-related challenges. By addressing these challenges, researchers can ensure the efficient utilization of resources and enhance the overall quality and impact of their research.

Funding Limitations

Securing adequate funding is often a significant challenge in clinical research. Limited funding can restrict the scope of research projects, hinder data collection and analysis, and limit the ability to recruit and retain qualified staff. Researchers should explore diverse funding sources, including governmental grants, philanthropic organizations, industry collaborations, and crowdfunding. Developing a well-designed research proposal, demonstrating the potential impact and significance of the study, and aligning the research objectives with funding agency priorities can increase the chances of securing funding. Additionally, establishing collaborations with other researchers or institutions can provide access to shared resources and enhance the competitiveness of funding applications.

Staff Allocation

Allocating and managing research staff efficiently is crucial for optimizing resource utilization. Researchers may face challenges in recruiting, training, and retaining qualified personnel, particularly in resource-constrained environments. It is important to carefully assess the required skills and expertise for each research position and develop job descriptions that align with the project's needs. Researchers should consider outsourcing certain tasks, collaborating with existing research networks or institutions, or leveraging volunteer resources where appropriate. Additionally, providing ongoing training and professional development opportunities for research staff can enhance their productivity and job satisfaction.

Infrastructure Requirements

Clinical research often requires access to specialized infrastructure, equipment, and facilities. Limited availability or inadequate infrastructure can pose challenges to data collection, analysis, and participant management. Researchers should identify the specific infrastructure requirements of their study and assess the availability of resources within their institution or research network. Collaboration with other institutions or research centers that possess the necessary infrastructure can help overcome these challenges. Moreover, exploring alternative solutions, such as shared resource centers or utilizing existing clinical settings, can help optimize resource utilization and mitigate infrastructure limitations.

Effective Budgeting

Developing a realistic and comprehensive budget is essential for effective resource management in clinical research. Researchers should carefully estimate the costs associated with study procedures, participant recruitment, data collection and analysis, and dissemination of findings. It is crucial to consider not only direct costs but also indirect costs, such as administrative expenses and overhead charges. Researchers should engage with financial experts or research administrators to ensure accurate budgeting and to navigate the specific funding requirements of different funding sources. Regular monitoring and tracking of expenses throughout the project duration can help identify any budget deviations and facilitate timely adjustments.

Project Management

Efficient project management plays a critical role in optimizing resource utilization and ensuring the timely completion of clinical research projects. Researchers should develop a comprehensive project plan, including a detailed timeline, clear milestones, and defined responsibilities for each team member. Effective communication and collaboration among team members are essential to minimize delays, address challenges proactively, and ensure the smooth progression of the study. Utilizing project management tools and software can facilitate efficient coordination, task tracking, and documentation.

Collaboration and Partnerships

Collaborating with other researchers, institutions, or organizations can help overcome resource limitations and enhance the feasibility and impact of clinical research projects. Establishing partnerships can provide access to shared resources, expertise, and infrastructure. Collaborative research networks or consortiums offer opportunities for pooling resources, sharing knowledge, and conducting large-scale studies. Researchers should actively seek opportunities for collaboration, engage in multidisciplinary research teams, and participate in research networks relevant to their area of study.

Conclusion

Effective resource management is crucial for the successful implementation of clinical research projects. By addressing challenges related to funding limitations, staff allocation, infrastructure requirements, and budgeting, researchers can optimize resource utilization and enhance the quality and impact of their research. Collaboration and partnerships with other researchers or institutions can provide access to shared resources and expertise. Furthermore, adopting efficient project management practices, ongoing monitoring of resource utilization, and flexibility in adapting to potential challenges contribute to the successful implementation of clinical research projects and the generation of valuable scientific knowledge.

SECTION 4: REGULATORY AND ETHICAL CHALLENGES

Regulatory and Ethical Challenges

Clinical research is subject to a range of regulatory and ethical requirements to ensure the protection of research participants, maintain scientific integrity, and uphold public trust. In this section, we will explore the regulatory and ethical challenges that researchers may encounter and discuss strategies for managing these challenges throughout the research process.

Navigating Complex Regulatory Frameworks

Clinical research is governed by a complex web of regulations, guidelines, and policies at local, national, and international levels. Researchers must familiarize themselves with the specific regulatory frameworks relevant to their study, such as those enforced by regulatory agencies like the Food and Drug Administration (FDA) in the United States or the European Medicines Agency (EMA) in the European Union. Understanding the specific requirements, including those related to study approvals, informed consent, data protection, and adverse event reporting, is essential. Proactive engagement with regulatory bodies, attending relevant workshops or training programs, and seeking guidance from regulatory experts can help researchers navigate these complexities effectively.

Obtaining Necessary Approvals

Ethical and regulatory approvals are typically required before initiating a clinical research study. Researchers must prepare and submit comprehensive study protocols, informed consent documents, and relevant supporting materials to the appropriate institutional review boards (IRBs), research ethics committees (RECs), or regulatory authorities. The process of obtaining approvals can be time-consuming and may vary across different jurisdictions. Researchers should carefully adhere to the submission requirements, anticipate potential delays, and allow sufficient time for the review and approval process. Proactive communication with IRBs, RECs, and regulatory authorities can help address any concerns or questions and expedite the approval process.

Ensuring Compliance with Ethical Guidelines

Ethical considerations play a fundamental role in clinical research. Researchers must adhere to ethical principles, such as respect for autonomy, beneficence, non-maleficence, and justice, throughout the research process. This includes ensuring informed consent, protecting participant privacy and confidentiality, and maintaining scientific integrity. Researchers should carefully design their study protocols to address ethical considerations, seek input from ethics experts or committees during the planning phase, and implement appropriate safeguards to protect the rights and welfare of research participants. Regular ethics training for research personnel can help raise awareness and promote ethical conduct in research.

Managing Ethical Concerns

During the course of a clinical research study, researchers may encounter ethical concerns or dilemmas that require careful consideration and resolution. These concerns may include issues such as participant recruitment and retention, participant vulnerability, conflicts of interest, or unexpected adverse events. It is essential for researchers to establish clear lines of communication within their research team and with relevant ethics committees to promptly address and resolve ethical concerns. Seeking guidance from ethics experts or engaging in ethical consultations can provide valuable insights and help navigate complex ethical dilemmas.

Maintaining Regulatory and Ethical Compliance

Maintaining regulatory and ethical compliance is an ongoing responsibility throughout the research process. Researchers should conduct regular internal audits to ensure adherence to regulatory requirements, ethical guidelines, and standard operating procedures. Monitoring and reporting adverse events or protocol deviations in a timely manner is essential for maintaining compliance and participant safety. Compliance with data protection regulations, such as the Health Insurance Portability and Accountability Act (HIPAA) or the General Data Protection Regulation (GDPR), should be prioritized to safeguard participant privacy and confidentiality. Regular updates on regulatory and ethical guidelines, attending conferences or workshops focused on research compliance, and engaging with professional networks can assist researchers in staying informed and up to date with evolving regulatory and ethical requirements.

Conclusion

Navigating the regulatory and ethical landscape in clinical research can present significant challenges. By proactively addressing these challenges, researchers can ensure regulatory and ethical compliance, protect the rights and welfare of research participants, and maintain the integrity and credibility of their research findings. Effective communication with regulatory bodies, ethics committees, and research personnel is crucial for managing these challenges throughout the research process. By staying informed, seeking expert guidance when needed, and adopting best practices, researchers can navigate regulatory and ethical challenges successfully and contribute to the advancement of scientific knowledge in an ethically responsible manner.

SECTION 5: DATA MANAGEMENT AND ANALYSIS CHALLENGES

Data Management and Analysis Challenges

Data management and analysis play a critical role in clinical research, and researchers often encounter various challenges throughout these processes. In this section, we will delve into the challenges related to data management and analysis in clinical research and discuss strategies for overcoming them effectively.

Data Quality Assurance

Ensuring data quality is crucial for reliable and valid research findings. Researchers face challenges in maintaining data accuracy, completeness, and consistency. To address these challenges, it is essential to implement robust data quality assurance measures. This includes designing comprehensive data collection forms and protocols, providing training to research staff on data collection procedures, conducting regular data quality checks and audits, and implementing data validation processes. Incorporating data validation rules, range checks, and logical consistency checks can help identify and resolve data quality issues in a timely manner.

Data Security

Data security is a significant concern in clinical research, particularly when handling sensitive participant information. Researchers must navigate challenges related to data storage, transmission, and protection. It is crucial to establish secure data storage systems, implement encryption measures, and ensure restricted access to confidential data. Compliance with data protection regulations, such as HIPAA or GDPR, is essential to safeguard participant privacy. Researchers should develop data security protocols, train research personnel on data protection practices, and regularly update security measures to mitigate the risk of data breaches.

Complex Data Analysis Techniques

Clinical research often involves complex data analysis techniques, such as advanced statistical methodologies or machine learning algorithms. Researchers may face challenges in selecting the appropriate analysis methods, implementing them correctly, and interpreting the results accurately. It is essential to have a solid understanding of the underlying statistical concepts and seek guidance from biostatisticians or data analysts when necessary. Collaborating with experts in data analysis can help researchers overcome challenges related to complex statistical techniques and ensure accurate and meaningful interpretation of study findings.

Training Research Staff

Proper training of research staff is crucial to address data management and analysis challenges effectively. Researchers should provide comprehensive training on data collection procedures, data entry protocols, and data management practices. Training should include guidance on maintaining data quality, adhering to standard operating procedures, and following ethical and regulatory guidelines. Investing in ongoing training and professional development for research staff can enhance their skills and competencies in data management and analysis, thereby mitigating potential challenges.

Technological Tools and Resources

Leveraging technological tools and resources can significantly aid in overcoming data management and analysis challenges. Researchers should explore available data management systems, electronic data capture tools, and statistical software packages to streamline data collection, storage, and analysis processes. These tools can assist in data validation, automate data cleaning procedures, and provide efficient data analysis capabilities. Regularly updating software and utilizing emerging technologies can help researchers stay current and leverage the latest advancements in data management and analysis.

Conclusion

Data management and analysis present significant challenges in clinical research. By implementing robust data quality assurance measures, prioritizing data security, leveraging appropriate statistical methodologies, and utilizing technological tools, researchers can overcome these challenges effectively. Training research staff and fostering collaboration with biostatisticians and data analysts are crucial for successful data management and analysis. Ultimately, addressing data-related challenges ensures the integrity, reliability, and validity of research findings, contributing to the advancement of clinical knowledge and patient care.

SECTION 6: COLLABORATION AND COMMUNICATION CHALLENGES

Collaboration and Communication Challenges

Collaboration and communication are vital components of successful clinical research, but researchers often face various challenges in these areas. In this section, we will explore the challenges associated with collaboration and communication in clinical research and provide strategies for overcoming them effectively.

Interdisciplinary Collaborations

Clinical research often involves interdisciplinary collaborations, bringing together researchers from different fields, such as medicine, nursing, statistics, and social sciences. Challenges may arise due to differences in research perspectives, terminology, and methodologies. To address these challenges, researchers should foster a culture of respect and open-mindedness, valuing the expertise and contributions of each discipline. Regular meetings, brainstorming sessions, and interdisciplinary workshops can facilitate knowledge sharing and bridge the gaps between disciplines. Collaborative research planning, where researchers from various disciplines actively participate in study design, can lead to more comprehensive and holistic research projects.

Example: In a clinical trial studying the effectiveness of a new drug, collaboration between medical researchers, biostatisticians, and pharmaceutical scientists is essential. Each discipline brings unique insights, with medical researchers providing clinical expertise, biostatisticians ensuring rigorous study design and data analysis, and pharmaceutical scientists contributing their knowledge of drug mechanisms and formulation.

Stakeholder Engagement

Engaging stakeholders, including patients, caregivers, healthcare professionals, policymakers, and community organizations, is crucial for the success and relevance of clinical research. Challenges in stakeholder engagement may involve identifying and recruiting appropriate stakeholders, managing competing interests and expectations, and maintaining ongoing communication throughout the research process. Researchers should develop strategies for identifying relevant stakeholders and involving them in research planning, implementation, and dissemination. Regular stakeholder meetings, focus groups, and advisory boards can facilitate meaningful engagement and ensure that research addresses the needs and priorities of stakeholders.

Example: In a study on improving cancer survivorship, researchers can engage stakeholders such as cancer survivors, oncologists, nurses, patient advocacy groups, and policymakers. Involving these stakeholders in research planning and conducting regular meetings to seek their input can enhance the study's relevance, ensure patient-centered outcomes, and promote the translation of research findings into policy and practice.

Effective Communication of Research Findings

Effectively communicating research findings to different audiences is essential for maximizing the impact of clinical research. Challenges may include communicating complex scientific information in a clear and understandable manner, tailoring messages to diverse audiences, and disseminating findings through appropriate channels. Researchers should prioritize clear and concise communication, avoiding jargon and technical terms when presenting research findings. They should adapt communication strategies to suit the target audience, such as using lay language when communicating with patients or presenting technical details to fellow researchers. Leveraging various communication channels, such as scientific publications, conference presentations, public lectures, and social media platforms, can enhance the reach and accessibility of research findings.

Example: In a study investigating the impact of a new intervention for diabetes management, researchers can communicate their findings through a combination of scientific publications for the research community, patient education materials for healthcare providers and patients, and public lectures or media interviews to raise awareness among the general public. Tailoring the message and selecting the appropriate communication channels for each audience can maximize the dissemination and uptake of research findings.

Building Strong Relationships

Building strong relationships among research team members is crucial for effective collaboration and communication. Challenges may include conflicts of interest, personality differences, and communication barriers. Researchers should prioritize relationship building and create a supportive and inclusive research environment. Regular team meetings, team-building activities, and establishing clear roles and responsibilities can foster positive team dynamics. Active listening, open and respectful communication, and conflict resolution strategies can help address challenges and promote effective teamwork.

Example: In a multicenter clinical trial, researchers from different institutions and departments must collaborate effectively. Building strong relationships through regular communication, shared decision-making, and recognition of individual contributions can promote a collaborative and cohesive research team.

Leveraging Communication Technologies

In the era of digital communication, researchers can leverage technology to overcome collaboration and communication challenges. Utilizing collaborative online platforms, video conferencing tools, and project management software can facilitate real-time communication and enhance collaboration among geographically dispersed team members. These technologies also enable efficient data sharing, document version control, and task coordination, streamlining the research process.

Example: Researchers can use online collaboration platforms, such as project management tools like Asana or communication platforms like Slack, to facilitate team communication, document sharing, and task management. Video conferencing tools like Zoom or Microsoft Teams can enable virtual meetings and facilitate real-time discussions among team members in different locations.

Conclusion

Addressing collaboration and communication challenges in clinical research is crucial for the success of research projects. By fostering interdisciplinary collaborations, engaging stakeholders, adopting effective communication strategies, building strong relationships, and leveraging communication technologies, researchers can overcome these challenges and enhance the quality and impact of their research. Emphasizing transparency, openness, and inclusivity throughout the research process promotes effective collaboration, maximizes stakeholder engagement, and ensures the dissemination of research findings to diverse audiences.

SECTION 7: ADAPTING TO CHANGES AND OVERCOMING UNEXPECTED CHALLENGES

Adapting to Changes and Overcoming Unexpected Challenges

In clinical research, unforeseen challenges and changes are inevitable. Researchers must be adaptable and resilient to effectively navigate these situations and ensure the successful completion of their studies. In this final section, we will explore the importance of adaptability and resilience in overcoming unexpected challenges in clinical research and provide strategies for addressing such challenges.

Unforeseen Events

Unforeseen events, such as natural disasters, epidemics, or changes in healthcare policies, can significantly impact research projects. These events may disrupt data collection, affect participant recruitment and retention, or introduce new ethical and regulatory considerations. Researchers should proactively assess potential risks and develop contingency plans to mitigate the impact of unforeseen events. This may involve modifying study timelines, adjusting recruitment strategies, or seeking alternative data collection methods.

Maintaining open lines of communication with study participants, research team members, and relevant stakeholders is crucial for effective coordination and problem-solving during unexpected events.

Example: During the COVID-19 pandemic, many clinical research studies faced challenges due to lockdowns, restricted access to healthcare facilities, and concerns for participant safety. Researchers had to quickly adapt their protocols to include virtual visits, remote data collection methods, and implement additional safety measures to continue their studies. Proactive risk assessment, ongoing communication, and flexibility in study design were key in successfully navigating these unexpected challenges.

Protocol Deviations

Protocol deviations can occur due to various reasons, such as changes in participant eligibility criteria, modifications to study procedures, or unforeseen adverse events. Researchers should establish a systematic approach for addressing protocol deviations, including documenting deviations, assessing their impact on study outcomes, and considering appropriate corrective actions. Communication with the research ethics committee or institutional review board (IRB) is essential to ensure compliance and maintain the integrity of the study. Transparent reporting of protocol deviations in research publications is necessary to provide a comprehensive understanding of the study's conduct and limitations.

Example: In a clinical trial, unexpected adverse events may require modifications to the study protocol to ensure participant safety. Researchers must promptly report such events to the IRB, assess their impact on the study's validity, and determine if any protocol amendments are necessary. Flexibility in modifying the protocol, while adhering to ethical guidelines and regulatory requirements, is essential for addressing unforeseen challenges and maintaining participant safety.

Changing Research Landscapes

The research landscape is constantly evolving, with advancements in technology, changes in clinical practice, and emerging scientific knowledge. Researchers must stay abreast of these developments and be willing to embrace innovative approaches to address new challenges. This may involve incorporating new technologies, exploring novel study designs, or collaborating with experts from emerging fields. Continuous learning and professional development are crucial for researchers to adapt to changing research landscapes and leverage new opportunities for advancing their studies.

Example: With the advent of digital health technologies, researchers are increasingly utilizing wearable devices, mobile applications, and electronic health records for data collection. Embracing these technologies allows researchers to gather real-time data, enhance participant engagement, and streamline data management. By being proactive in adopting innovative approaches, researchers can stay ahead of the curve and overcome potential challenges in integrating new technologies into their studies.

Proactive Risk Assessment and Contingency Planning

To navigate unexpected challenges effectively, researchers should conduct proactive risk assessments during the study planning phase. Identifying potential risks and developing contingency plans can help mitigate the impact of unforeseen events or deviations from the protocol. Researchers should anticipate challenges related to participant recruitment, data quality, regulatory compliance, and resource limitations. By addressing these challenges proactively, researchers can minimize disruptions, ensure participant safety, and maintain the scientific rigor of their studies.

Example: Before initiating a clinical trial, researchers can conduct a thorough risk assessment, considering factors such as recruitment difficulties, unforeseen adverse events, or changes in study logistics. Based on this assessment, they can develop contingency plans, such as alternative recruitment strategies, additional monitoring procedures, or flexible data collection methods. Proactive risk assessment and contingency planning enable researchers to respond effectively to unexpected challenges and maintain the integrity of the study.

Conclusion

Adaptability and resilience are essential qualities for researchers to successfully overcome unexpected challenges in clinical research. By proactively assessing risks, developing contingency plans, embracing innovative approaches, and maintaining open communication with stakeholders, researchers can navigate unforeseen events, protocol deviations, and changing research landscapes. Continuous learning, flexibility in study design, and a proactive problem-solving mindset are key to addressing challenges and ensuring the successful completion of clinical research projects.

By the end of Chapter 10, readers will have a comprehensive understanding of the common challenges encountered in clinical research and effective strategies for managing and overcoming them. They will be equipped with the knowledge and skills necessary to proactively identify and address challenges, optimize resource allocation, navigate regulatory and ethical complexities, manage data effectively, foster collaboration and communication, and adapt to unexpected circumstances. This knowledge will empower researchers to successfully navigate the complex landscape of clinical research and maximize the chances of research success.

11

TRANSLATING RESEARCH INTO PRACTICE: IMPLEMENTATION AND IMPACT

CHAPTER XI

TRANSLATING RESEARCH INTO PRACTICE: IMPLEMENTATION AND IMPACT

Chapter 11 delves into the intricate process of translating research into practice, emphasizing the implementation and impact of clinical research findings. This chapter explores the concept of translational research and its importance in bridging the gap between scientific discoveries and their practical applications in healthcare settings. By examining various strategies and frameworks for implementation science, the chapter provides insights into the development, dissemination, and adoption of clinical guidelines and protocols. Furthermore, it addresses the challenges encountered in translating research findings into practice, offering solutions for overcoming barriers and facilitating change. Through a comprehensive understanding of these processes, readers will be equipped to effectively implement evidence-based practices, ensuring that research findings have a meaningful impact on patient care and healthcare outcomes.

SECTION 1: INTRODUCTION TO TRANSLATIONAL RESEARCH

Introduction to Translational Research

In this section, we will provide a thorough introduction to the concept of translational research and its significance in bridging the gap between laboratory discoveries and clinical applications. Translational research plays a critical role in accelerating the translation of scientific knowledge into tangible benefits for patients, ultimately improving healthcare outcomes. It encompasses a multidisciplinary approach that integrates basic science, preclinical studies, and clinical trials to facilitate the development of new diagnostics, treatments, and interventions.

Translational research serves as a bridge between the "bench" (where laboratory research is conducted) and the "bedside" (where clinical care is delivered). It aims to translate fundamental discoveries made at the laboratory bench into practical applications at the patient's bedside, thereby improving healthcare delivery and patient outcomes. By facilitating the efficient translation of scientific discoveries, translational research helps to address unmet medical needs and bring innovative solutions to the forefront of clinical practice.

The translational research continuum involves multiple stages that collectively contribute to the successful translation of scientific knowledge. It begins with basic science research, where scientists explore fundamental biological mechanisms, identify potential targets, and generate hypotheses. Basic science research provides the foundation for subsequent stages of translational research.

Preclinical studies form the next phase of translational research, where scientists investigate the safety, efficacy, and mechanism of action of potential interventions using laboratory models, cell cultures, and animal models. These studies help researchers gain insights into the potential therapeutic effects of new interventions and provide a basis for making informed decisions regarding their progression to clinical trials.

Clinical trials represent the final stage of the translational research continuum. These trials involve the systematic evaluation of new diagnostics, treatments, or interventions in human subjects to determine their safety, efficacy, and effectiveness. Clinical trials play a pivotal role in generating evidence that informs clinical practice and regulatory decision-making.

The translational research continuum consists of different stages that encompass the progression of research from the laboratory bench to clinical settings and eventually to public health and policy implementation. These stages include:

T1: Basic Science to Clinical Research: This stage focuses on translating basic scientific discoveries into clinical research and interventions. It involves conducting preclinical studies, including animal and in vitro experiments, to assess the feasibility, safety, and effectiveness of potential interventions. Examples include testing a novel drug candidate in animal models or conducting early-phase clinical trials in humans.

T2: Clinical Research to Clinical Practice: At this stage, research findings from clinical studies are translated into clinical practice guidelines and evidence-based interventions. It involves conducting randomized controlled trials (RCTs) and observational studies to evaluate the effectiveness and safety of interventions in larger patient populations. Examples include conducting multi-center RCTs to assess the efficacy of a new treatment approach or developing clinical practice guidelines based on systematic reviews and meta-analyses.

T3: Clinical Practice to Health Outcomes: This stage focuses on evaluating the impact of interventions on health outcomes in real-world settings. It involves implementing evidence-based interventions in clinical practice and assessing their effectiveness, patient adherence, and long-term outcomes. Examples include studying the effectiveness of a new treatment protocol in routine clinical care or evaluating the impact of a public health intervention on population health outcomes.

T4: Health Outcomes to Population Health Impact: The final stage of the translational research continuum involves translating research findings into population health strategies, policies, and public health interventions. It aims to address broader health disparities, promote health equity, and inform healthcare policies and resource allocation. Examples include using research evidence to advocate for policy changes, implementing public health campaigns based on research findings, or informing healthcare guidelines at a population level

Translational research is of paramount importance as it has the potential to directly impact patient care, inform healthcare decision-making, and improve health outcomes. By translating research findings into practical applications, translational research ensures that scientific discoveries are effectively applied in clinical settings, leading to evidence-based practice. It also facilitates the development of new therapies, interventions, and healthcare policies that can address unmet clinical needs and improve healthcare delivery and outcomes.

Collaboration is a fundamental aspect of translational research. Successful translation requires close collaboration among researchers, clinicians, industry partners, and regulatory agencies. Collaboration fosters the exchange of knowledge, resources, and expertise, and enables the efficient progression of research from the laboratory to clinical practice. For example, researchers working in academic institutions collaborate with clinicians to identify clinical needs, while industry partners contribute their expertise in drug development, manufacturing, and commercialization.

Putting It All Together

Examples of translational research proliferate in various fields. One notable example is the development of targeted therapies for cancer treatment. Basic science research identified specific genetic mutations driving certain cancers, leading to the development of targeted drugs that inhibit these mutated genes or proteins. Through preclinical studies and subsequent clinical trials, these targeted therapies have transformed cancer treatment, offering more effective and personalized approaches with improved patient outcomes.

The development of targeted therapies for cancer exemplifies the power of translational research in revolutionizing cancer treatment. This example highlights the progression from fundamental discoveries at the laboratory bench to clinical applications that have significantly improved patient outcomes.

The story begins with basic science research, where scientists identified specific genetic mutations or alterations in cancer cells that drive the development and progression of certain types of cancer. For instance, in chronic myeloid leukemia (CML), researchers discovered a chromosomal translocation that results in the formation of the Philadelphia chromosome and the production of a mutated protein called BCR-ABL. This genetic abnormality was found to be a key driver of CML.

Building upon this fundamental knowledge, researchers embarked on preclinical studies to explore the therapeutic potential of targeting the BCR-ABL protein. Using laboratory models, such as cell cultures and animal models, scientists developed compounds that could inhibit the activity of BCR-ABL and selectively target cancer cells while sparing healthy cells. These preclinical studies provided promising evidence that targeting BCR-ABL could effectively suppress cancer growth and potentially lead to improved treatment outcomes for CML patients.

The positive findings from preclinical studies paved the way for clinical trials, where the safety, efficacy, and effectiveness of these targeted therapies were rigorously evaluated in human subjects. Clinical trials typically follow a phased approach, starting with Phase I trials to assess safety and dosage, followed by Phase II trials to evaluate efficacy in a larger patient population, and concluding with Phase III trials to compare the new therapy with standard treatments.

In the case of targeted therapies for CML, clinical trials involving patients with CML were conducted to assess the effectiveness of BCR-ABL inhibitors. These trials demonstrated remarkable responses in terms of reducing cancer burden, improving patient outcomes, and even achieving complete remission in some cases. The targeted therapies not only showed increased effectiveness but also exhibited reduced toxicity compared to conventional chemotherapy.

As the clinical trial data accumulated, regulatory agencies such as the U.S. Food and Drug Administration (FDA) evaluated the safety and efficacy profiles of these targeted therapies. Based on the evidence generated from the clinical trials, the regulatory agencies granted approval for the use of targeted therapies in the treatment of CML.

These therapies, such as imatinib (Gleevec), dasatinib (Sprycel), and nilotinib (Tasigna), have become standard of care for CML patients and have significantly improved their prognosis and quality of life.

The success story of targeted therapies for CML exemplifies the translational research process in action. Basic science research elucidated the underlying genetic mechanisms driving cancer, leading to the development of targeted therapies through preclinical studies. These therapies were subsequently tested in clinical trials, where their safety and efficacy were rigorously evaluated in patients. The collective findings from basic science, preclinical studies, and clinical trials provided the necessary evidence to support the approval and widespread adoption of these targeted therapies in clinical practice.

This example underscores the importance of translational research in transforming scientific discoveries into tangible clinical applications. Through collaboration among researchers, clinicians, industry partners, and regulatory agencies, targeted therapies emerged as a paradigm-shifting approach in cancer treatment. The success of targeted therapies for CML serves as a testament to the power of translational research in improving patient outcomes and illustrates the impact that collaborative efforts across the translational research continuum can have on advancing medical science.

In conclusion, translational research plays a vital role in bridging the gap between laboratory discoveries and clinical applications. By facilitating the translation of scientific knowledge into practical solutions, translational research accelerates the development of new diagnostics, treatments, and interventions. Collaboration among researchers, clinicians, industry partners, and regulatory agencies is essential to ensure the successful translation of scientific discoveries into improved patient outcomes. The translational research continuum, encompassing basic science research, preclinical studies, and clinical trials, serves as a roadmap for advancing scientific discoveries from the bench to the bedside.

SECTION 2: FROM BENCH TO BEDSIDE: BASIC SCIENCE RESEARCH

From Bench to Bedside: Basic Science Research

In this section we will delve into the role of basic science research in translational research, focusing on its importance in understanding disease mechanisms, identifying therapeutic targets, and developing novel interventions. We will explore the process of hypothesis generation, experimental design, data collection, and analysis in basic science research. Additionally, we will discuss strategies for enhancing the translation of basic science findings into clinically relevant applications, including collaboration with clinicians, patient-oriented research, and the identification of biomarkers for early detection and personalized medicine.

Basic science research forms the foundation of translational research by investigating fundamental biological processes and elucidating the underlying mechanisms of diseases. This type of research often occurs in laboratory settings and involves studying cells, molecules, and animal models to unravel the intricacies of diseases at the molecular and cellular levels.

Hypothesis generation is a critical step in basic science research. Scientists develop hypotheses based on existing knowledge, observations, and gaps in understanding. For example, if researchers observe aberrant cell growth in cancer, they may hypothesize that a specific signaling pathway is dysregulated and contributes to tumor formation. These hypotheses drive the experimental design and guide subsequent investigations.

Experimental design in basic science research involves planning and implementing experiments to test hypotheses and gather relevant data. Researchers carefully select appropriate model systems, such as cell lines, animal models, or even human tissue samples, to mimic the disease or process of interest. They design experiments with controls to ensure reliable comparisons and employ techniques and technologies suitable for studying the specific research question.

Data collection in basic science research involves conducting experiments and gathering empirical evidence. Scientists perform various assays, measurements, and observations to obtain data related to their research objectives. For instance, in the context of cancer research, data collection may involve measuring gene expression levels, analyzing protein-protein interactions, or characterizing cellular responses to stimuli.

Once the data is collected, researchers analyze and interpret it using statistical methods and data visualization tools. This analysis helps identify patterns, relationships, and statistical significance. The findings are then compared to existing knowledge and used to draw conclusions and refine hypotheses. Importantly, basic science research often generates large amounts of data, necessitating robust data management and analysis techniques.

To enhance the translation of basic science findings into clinically relevant applications, collaboration with clinicians and other stakeholders is crucial. By engaging with clinicians, basic scientists can gain insights into the clinical aspects of the disease and the challenges faced in patient care. This collaboration helps align basic science research with the needs and priorities of clinical practice, ensuring that the research is relevant and addresses clinically important questions.

Patient-oriented research is another valuable strategy for translating basic science findings. This approach involves actively involving patients in the research process, such as participating in clinical trials, providing samples for analysis, or contributing to the design and implementation of research studies. Patient-oriented research promotes patient-centered outcomes and helps bridge the gap between basic science discoveries and their application in real-world clinical settings.

Furthermore, identifying biomarkers plays a significant role in the translation of basic science findings. Biomarkers are measurable indicators that provide information about a biological state, condition, or response to therapy. By identifying biomarkers associated with a particular disease, researchers can develop diagnostic tests, prognostic tools, and personalized treatment approaches. For example, in cancer research, the identification of specific genetic mutations or protein markers has led to the development of targeted therapies tailored to individual patients.

One example of basic science research contributing to clinical applications is the discovery of human papillomavirus (HPV) as a major cause of cervical cancer. Through basic science research, scientists identified the role of specific HPV types in the development of cervical cancer and uncovered the molecular mechanisms involved. This knowledge led to the development of HPV vaccines, which have significantly reduced the incidence of cervical cancer worldwide.

In summary, basic science research plays a pivotal role in translational research by providing a deep understanding of disease mechanisms, identifying therapeutic targets, and facilitating the development of novel interventions. Through hypothesis generation, experimental design, data collection, and analysis, basic scientists contribute essential insights to the field. By collaborating with clinicians, involving patients in research, and identifying biomarkers, basic science findings can be effectively translated into clinically relevant applications, improving patient care and outcomes.

SECTION 3: PRECLINICAL STUDIES: BRIDGING THE GAP

Preclinical Studies: Bridging the Gap

Next up, we will explore the importance of preclinical studies in translational research. We will discuss the role of preclinical models, including animal models and in vitro studies, in evaluating the safety and efficacy of potential interventions before testing in humans. Readers will gain an understanding of the regulatory requirements, ethical considerations, and best practices for conducting preclinical studies. Additionally, we will explore strategies for improving the translational potential of preclinical findings, such as refining study design, incorporating clinically relevant endpoints, and promoting transparency in reporting preclinical results.

Preclinical studies are a critical component of translational research, serving as a bridge between basic science research and clinical trials. These studies involve evaluating the safety, efficacy, and mechanisms of action of potential interventions, such as new drugs, therapies, or medical devices, before they are tested in human subjects. Preclinical studies are typically conducted using various models, including animal models and in vitro systems, to mimic the disease conditions and evaluate the potential benefits and risks of the intervention.

Animal models are frequently used in preclinical studies to assess the safety and effectiveness of potential treatments. These models can replicate certain aspects of human physiology, pathology, and disease progression, providing valuable insights into the intervention's potential impact. For example, in cancer research, animal models can be used to evaluate the efficacy of new anticancer drugs, study tumor growth and metastasis, and assess potential toxicities.

In vitro studies, which involve conducting experiments using isolated cells or tissues outside of living organisms, also play a crucial role in preclinical research. In vitro models allow researchers to investigate specific cellular mechanisms, test the effects of interventions on cell cultures, and gain a better understanding of the intervention's biological effects. These studies provide valuable preliminary data and help guide the decision-making process before proceeding to animal studies and human trials.

When conducting preclinical studies, researchers must adhere to regulatory requirements and ethical considerations. Regulatory bodies, such as the Food and Drug Administration (FDA) in the United States, provide guidelines and oversight to ensure the safety and scientific rigor of preclinical research. Ethical considerations include minimizing harm to animals, following animal welfare regulations, and using the minimum number of animals necessary to achieve valid results. Researchers must also consider the principles of the 3Rs: Replacement, Reduction, and Refinement, which promote the use of alternative methods, reducing the number of animals used, and refining experimental techniques to minimize animal suffering.

To enhance the translational potential of preclinical findings, several strategies can be employed. First, refining study design is crucial to ensure that preclinical studies are conducted in a way that closely reflects the intended clinical applications. For example, using relevant disease models, considering appropriate dosing regimens, and incorporating relevant endpoints that align with clinical outcomes are essential aspects of study design.

Incorporating clinically relevant endpoints is another important consideration in preclinical studies. Endpoints are measurable outcomes used to evaluate the effects of the intervention. By selecting endpoints that are meaningful in the context of human diseases, such as tumor regression, survival rates, or functional improvements, researchers can better assess the potential clinical impact of the intervention.

Promoting transparency in reporting preclinical results is vital for ensuring the reproducibility and reliability of the findings. Transparent reporting allows other researchers to evaluate and validate the results, facilitating the translation of preclinical findings into clinical applications. Initiatives such as the ARRIVE guidelines (Animal Research: Reporting of In Vivo Experiments) provide recommendations for reporting preclinical research, including study design, experimental procedures, and statistical analyses.

One example of the importance of preclinical studies in translational research is the development of targeted therapies for various diseases. Preclinical studies involving animal models and in vitro systems have been instrumental in identifying potential therapeutic targets, evaluating the efficacy of targeted drugs, and understanding their mechanisms of action. For instance, in the field of cancer research, preclinical studies have contributed to the development of targeted therapies that specifically inhibit certain molecular pathways involved in tumor growth and survival. These therapies have shown promising results in preclinical studies, leading to further evaluation in clinical trials and eventually benefiting patients by providing more effective and less toxic treatment options.

In summary, preclinical studies serve as a crucial link between basic science research and clinical trials in translational research. They provide valuable insights into the safety and efficacy of potential interventions before testing in humans. By using appropriate models, adhering to regulatory and ethical considerations, refining study design, incorporating clinically relevant endpoints, and promoting transparent reporting, preclinical studies can enhance the translational potential of research findings and contribute to the development of novel treatments and interventions for various diseases.

SECTION 4: CLINICAL TRIALS: FROM BEDSIDE TO PRACTICE

Clinical Trials: From Bedside to Practice

In the section, we will continue the discussion into the pivotal role of clinical trials in translational research. We will discuss the different phases of clinical trials, including safety and efficacy assessments, and emphasize the importance of rigorous study design, participant recruitment, and data collection. Readers will gain insights into the regulatory requirements, ethical considerations, and best practices for conducting clinical trials. Additionally, we will explore strategies for enhancing the translation of clinical trial findings into clinical practice, such as patient-centered outcomes research, adaptive trial designs, and evidence-based medicine.

Clinical trials are a critical component of translational research, serving as the final step in evaluating the safety and effectiveness of new interventions before they can be implemented into clinical practice. These trials are conducted with human participants and are designed to generate evidence about the intervention's efficacy, side effects, optimal dosages, and potential interactions with other treatments.

Clinical trials typically progress through several phases, each with specific objectives and increasing levels of evidence. Phase I trials focus on assessing the intervention's safety and dosing, usually involving a small number of healthy volunteers or patients. Phase II trials involve a larger number of participants and aim to evaluate the intervention's efficacy and side effects. Phase III trials, often involving thousands of participants, compare the intervention to existing standard treatments or placebos to determine its effectiveness and further assess safety. Finally, Phase IV trials, conducted after regulatory approval, continue to monitor the intervention's long-term effects and gather additional real-world data.

Rigorous study design is crucial in clinical trials to ensure reliable and meaningful results. Randomized controlled trials (RCTs) are considered the gold standard design, where participants are randomly assigned to either the intervention group or the control group. This design helps minimize bias and provides a basis for comparing the intervention's effects to a control or standard treatment. Blinding, whether single-blind (participants unaware of the treatment received) or double-blind (both participants and researchers unaware), further enhances the objectivity of the trial.

Participant recruitment is a critical aspect of clinical trials, as adequate representation of the target population is necessary to ensure the generalizability of the findings. Recruitment strategies may involve collaboration with healthcare providers, patient advocacy groups, and community outreach programs. Ensuring diverse participation in clinical trials is particularly important to assess how interventions may affect different populations, including underrepresented groups.

Ethical considerations play a central role in clinical trials. Institutional review boards (IRBs) or ethics committees provide oversight and evaluate the ethical soundness of the trial protocol, participant informed consent processes, and the balance between potential benefits and risks. Informed consent, voluntary participation, confidentiality, and participant welfare are essential ethical principles that must be upheld throughout the trial.

To enhance the translation of clinical trial findings into clinical practice, several strategies can be employed. Patient-centered outcomes research aims to align research with patient preferences, values, and goals. This approach involves engaging patients and caregivers in trial design, identifying outcomes that matter most to them, and evaluating interventions from a patient-centered perspective.

Adaptive trial designs are another strategy to enhance translation. These designs allow for modifications to the trial protocol based on interim data analysis, enabling more efficient decision-making and potential modifications to the intervention or the trial's conduct. Adaptive designs can help reduce sample sizes, decrease trial duration, and increase the likelihood of identifying effective treatments.

Evidence-based medicine, which integrates clinical expertise, patient values, and the best available research evidence, is a vital approach to ensure that clinical trial findings are applied in practice. Systematic reviews and meta-analyses, which synthesize the results of multiple trials, provide a comprehensive and unbiased evaluation of the evidence. Clinical practice guidelines, developed based on the best available evidence, help clinicians make informed decisions about the most effective interventions for specific patient populations or conditions.

Let's consider an example to illustrate the impact of clinical trials in translational research. Suppose a novel drug targeting a specific cancer pathway shows promising results in preclinical studies and early-phase clinical trials. As the drug progresses to Phase III trials, researchers conduct a large-scale, multicenter trial involving hundreds or thousands of patients with the target cancer. This trial employs a randomized controlled design, where participants are assigned either the new drug or the standard treatment. The trial measures various endpoints, such as overall survival, progression-free survival, and quality of life.

Upon completion of the trial, data analysis reveals that patients receiving the new drug demonstrate significantly improved outcomes compared to those receiving the standard treatment. These findings support the efficacy and safety of the new drug in a large and diverse patient population. Regulatory agencies review the trial results, and if the benefits outweigh the risks, the drug may receive regulatory approval for clinical use.

Once approved, the findings from the clinical trial can guide clinical practice, enabling healthcare providers to offer the new drug as a treatment option to patients with the specific cancer type. The trial results may also influence treatment guidelines and protocols, leading to broader adoption of the intervention within the medical community.

In summary, clinical trials play a crucial role in translational research by providing evidence of the safety, efficacy, and effectiveness of new interventions before they are implemented in clinical practice. Rigorous study design, participant recruitment, and ethical considerations are vital aspects of conducting clinical trials. Strategies such as patient-centered outcomes research, adaptive trial designs, and evidence-based medicine help bridge the gap between trial findings and clinical practice, ensuring that patients receive the most effective and evidence-based treatments.

SECTION 5: IMPLEMENTATION SCIENCE: MOVING FROM EVIDENCE TO PRACTICE

Implementation Science: Moving from Evidence to Practice

In this section, we will explore the field of implementation science and its crucial role in translational research. We will discuss the importance of implementation science in bridging the gap between research evidence and routine clinical practice. Readers will gain an understanding of implementation strategies, such as dissemination and implementation research, knowledge translation, and quality improvement initiatives. Additionally, we will delve into the challenges and considerations involved in implementing evidence-based practices in real-world healthcare settings, including organizational barriers, stakeholder engagement, and sustainability.

Implementation science focuses on understanding how to effectively and efficiently translate research evidence into routine clinical practice to improve patient outcomes and population health. It involves studying the methods, processes, and factors that influence the successful adoption, integration, and sustainability of evidence-based interventions in diverse healthcare settings.

Dissemination and implementation research plays a central role in implementation science. Dissemination refers to the systematic distribution of research findings and information to the target audience, while implementation focuses on the strategies and approaches to integrate evidence-based interventions into routine care. Dissemination and implementation research aim to identify and address barriers and facilitators to the uptake of evidence-based practices, with a focus on understanding the context in which implementation occurs.

Knowledge translation is another key concept in implementation science. It involves the synthesis, exchange, and application of research findings to inform healthcare decision-making. Knowledge translation aims to ensure that research evidence is effectively communicated and used by healthcare providers, policymakers, and other stakeholders to improve health outcomes. This process involves engaging stakeholders throughout the research process, identifying their needs and preferences, and tailoring the dissemination and implementation strategies accordingly.

Quality improvement initiatives are an essential component of implementation science. These initiatives focus on systematically improving the delivery of healthcare services and integrating evidence-based practices into routine care. Quality improvement efforts involve assessing current practices, identifying gaps, implementing changes, and continuously monitoring and evaluating the impact of these changes on patient outcomes. Examples of quality improvement approaches include the Plan-Do-Study-Act (PDSA) cycles, Lean Six Sigma methodologies, and the use of improvement collaboratives or learning networks.

Implementing evidence-based practices in real-world healthcare settings poses several challenges. Organizational barriers, such as resistance to change, lack of resources, and competing priorities, can hinder the adoption and sustainability of interventions. Engaging stakeholders, including healthcare providers, administrators, patients, and policymakers, is critical for successful implementation. Collaboration and effective communication among stakeholders help build support, address concerns, and foster a shared vision of implementing evidence-based practices.

Sustainability is another key consideration in implementation science. It involves ensuring that implemented interventions are maintained over time and become integrated into routine practice. Strategies for sustainability may include ongoing training and education, monitoring and feedback mechanisms, and creating a culture of continuous quality improvement. Long-term financial and organizational support are also essential for sustaining evidence-based practices beyond the initial implementation phase.

An example of implementation science in action is the adoption of a new clinical pathway for managing a chronic disease, such as diabetes. Based on research evidence, a new guideline is developed that outlines best practices for diabetes management. Implementation scientists would then work to disseminate this guideline to healthcare providers and develop strategies to facilitate its integration into routine care. This may involve conducting training workshops, providing decision support tools, and fostering collaboration among providers to share experiences and learn from each other. Ongoing monitoring and evaluation of the implementation process would help identify barriers, address challenges, and refine the approach to improve the uptake and sustainability of the new clinical pathway.

In summary, implementation science plays a crucial role in translational research by focusing on the effective translation of research evidence into routine clinical practice. Dissemination and implementation research, knowledge translation, and quality improvement initiatives are key strategies in implementation science.

Overcoming organizational barriers, engaging stakeholders, and ensuring sustainability are critical considerations in implementing evidence-based practices in real-world healthcare settings. Through implementation science, researchers and practitioners can bridge the gap between research evidence and practice, leading to improved patient outcomes and population health.

SECTION 6: REGULATORY AND ETHICAL CONSIDERATIONS IN TRANSLATIONAL RESEARCH

Regulatory and Ethical Considerations in Translational Research

In this section, we will focus on the crucial role of regulatory agencies and ethical considerations in ensuring the safety, efficacy, and ethical conduct of translational research. We will discuss the role of regulatory agencies, such as the Food and Drug Administration (FDA) in the United States, in overseeing the development and approval of new interventions. Additionally, we will explore the regulatory pathways for drug development, investigational device exemptions, and ethical considerations in human subjects research. Strategies for navigating regulatory and ethical challenges, including informed consent, protection of human subjects, and adherence to Good Clinical Practice (GCP) guidelines, will be discussed.

Regulatory agencies play a vital role in translational research by ensuring that new interventions, such as drugs, biologics, and medical devices, are safe and effective for use in humans. In the United States, the FDA is responsible for regulating the development, approval, and post-marketing surveillance of medical products. The

FDA reviews preclinical and clinical data to evaluate the safety and efficacy of new interventions before granting marketing approval.

The regulatory pathway for drug development typically involves several phases, starting with preclinical studies in the laboratory and animal models, followed by clinical trials in humans. The FDA reviews the data from these studies to assess the intervention's safety and effectiveness, as well as its benefits and risks. The FDA also evaluates the manufacturing processes, labeling, and post-marketing surveillance plans to ensure ongoing safety monitoring.

Similarly, investigational device exemptions (IDE) are required for the conduct of clinical investigations involving medical devices. The IDE process involves submitting an application to the FDA, providing detailed information on the device, the proposed study design, and the anticipated risks and benefits. The FDA reviews the IDE application to assess the device's safety and the scientific validity of the proposed study.

Ethical considerations are paramount in translational research, particularly in studies involving human subjects. Protecting the rights, welfare, and well-being of participants is of utmost importance. Researchers must obtain informed consent from participants, ensuring they understand the study purpose, procedures, potential risks and benefits, and their rights to withdraw at any time without consequences. Informed consent is a process that involves ongoing communication and ensures participants' voluntary and informed participation.

Protection of human subjects goes beyond informed consent. Institutional Review Boards (IRBs) or Ethics Committees play a crucial role in reviewing and approving research protocols to ensure ethical standards are met. IRBs assess the study design, participant selection criteria, data collection and management procedures, and plans for monitoring participant safety. They also review the potential conflicts of interest and financial disclosures of the researchers.

Adherence to Good Clinical Practice (GCP) guidelines is essential for conducting ethical and scientifically sound clinical research. GCP provides a framework for the design, conduct, and reporting of clinical trials. It includes guidelines for study documentation, data collection and monitoring, participant safety, and study drug accountability. Adhering to GCP principles ensures the integrity and reliability of research data and protects the rights and well-being of study participants.

Navigating regulatory and ethical challenges in translational research requires careful planning and adherence to established guidelines and regulations. Researchers should establish strong collaborations with regulatory professionals, seek guidance from institutional review boards, and consult regulatory agencies to ensure compliance with applicable regulations and ethical standards. Regular training and education on regulatory and ethical principles are also essential to maintain a high level of awareness and understanding among researchers and study staff.

Example 1: In a translational research study involving the development of a novel cancer therapy, the researchers would need to comply with regulatory requirements. They would conduct preclinical studies in the laboratory and animal models to gather data on the intervention's safety and efficacy. The researchers would submit an Investigational New Drug (IND) application to the FDA, providing comprehensive data on the intervention's preclinical results, manufacturing processes, and proposed clinical trial design. The FDA would review the IND application, and if approved, the researchers would proceed with clinical trials involving human participants, following the FDA's regulations and guidelines for good clinical practice.

Example 2: In a clinical trial testing a new medical device, researchers would need to obtain informed consent from the participants before their enrollment. The informed consent process would involve providing detailed information about the study, including the purpose, procedures, potential risks and benefits, and the participants' rights. The researchers would submit the study protocol and informed consent documents to the institutional review board for review and approval. The IRB would assess the ethical considerations, such as participant protection, data privacy, and study monitoring. Throughout the trial, the researchers would adhere to Good Clinical Practice guidelines, ensuring the data collected is reliable and the participants' rights and safety are safeguarded.

By adhering to regulatory requirements and ethical considerations, researchers can ensure the integrity of their translational research and protect the rights and well-being of study participants. Collaboration with regulatory agencies, institutional review boards, and ethics committees is essential to navigate the complex landscape of regulatory and ethical considerations in translational research.

SECTION 7: FUTURE DIRECTIONS IN TRANSLATIONAL RESEARCH

Future Directions in Translational Research

In this final section, we will discuss emerging trends and potential avenues for advancing translational research efforts. We will explore innovative technologies, interdisciplinary collaboration, big data analytics, real-time data monitoring, patient-centered research, community engagement, and global health initiatives as key elements shaping the future of translational research.

Innovative Technologies

Genomics, precision medicine, and digital health are revolutionizing the field of translational research. Genomic sequencing and analysis provide insights into the genetic basis of diseases, enabling personalized treatment approaches. Precision medicine tailors interventions based on an individual's genetic, environmental, and lifestyle factors. Digital health technologies, such as wearables, mobile apps, and remote monitoring devices, facilitate real-time data collection, patient engagement, and personalized interventions.

Example: In cancer research, genomic profiling helps identify specific genetic alterations driving tumor growth. This information can guide the selection of targeted therapies tailored to the individual's genetic profile, improving treatment outcomes and minimizing adverse effects.

Interdisciplinary Collaboration

Collaboration across disciplines, such as medicine, biology, engineering, data science, and behavioral science, is crucial for addressing complex health challenges. Translational research benefits from diverse perspectives, expertise, and innovative approaches offered by interdisciplinary teams. Collaboration among researchers, clinicians, industry partners, and patients accelerates the translation of research findings into clinical practice.

Example: A collaborative team consisting of researchers, clinicians, and engineers may work together to develop and refine medical devices or technologies, combining their knowledge and skills to improve patient outcomes and treatment effectiveness.

Big Data Analytics

The availability of large-scale datasets, often referred to as big data, presents opportunities for extracting valuable insights in translational research. Advanced analytics, including machine learning and artificial intelligence, can help identify patterns, predict treatment responses, and guide decision-making. Integrating diverse datasets, such as electronic health records, genomic data, and social determinants of health, can provide a comprehensive understanding of diseases and facilitate personalized interventions.

Example: Analyzing vast amounts of patient data can help identify patterns and risk factors associated with certain diseases, enabling early detection and targeted interventions. Machine learning algorithms can assist in predicting patient responses to specific treatments, allowing for personalized therapeutic approaches.

Real-Time Data Monitoring

Continuous real-time data monitoring, enabled by wearable devices, remote sensors, and mobile health applications, offers opportunities for precise and timely monitoring of patient health and treatment responses. Real-time data collection provides a dynamic and comprehensive view of patients' health status, facilitating early detection of adverse events, treatment optimization, and personalized interventions.

Example: Remote monitoring devices can track vital signs, medication adherence, and activity levels in patients with chronic conditions. Real-time data alerts healthcare providers to any deviations from normal values, enabling timely interventions and reducing hospitalizations.

Patient-Centered Research

Emphasizing patient perspectives, needs, and values in the research process is crucial for effective translational research. Engaging patients as partners in research design, implementation, and dissemination ensures that research findings are relevant, acceptable, and applicable to the target population. Patient-centered research promotes patient empowerment, shared decision-making, and improved healthcare outcomes.

Example: Involving patients in the design of clinical trials ensures that research outcomes align with patients' priorities, minimizes the burden of participation, and enhances the relevance and generalizability of the results.

Community Engagement and Global Health Initiatives

Engaging communities and addressing health disparities are essential in translational research. Collaborating with community organizations, healthcare providers, and policymakers fosters the implementation of research findings in real-world settings. Global health initiatives aim to address global health challenges, promote equity, and improve health outcomes across diverse populations.

Example: A research study may partner with local community centers or clinics to conduct health screenings, provide education, and facilitate access to care for underserved populations. These partnerships empower communities, enhance research participation, and increase the impact of translational research.

Conclusion

In summary, future directions in translational research involve leveraging innovative technologies, fostering interdisciplinary collaboration, harnessing the power of big data analytics, implementing real-time data monitoring, prioritizing patient-centered research, engaging communities, and addressing global health challenges. By embracing these advancements, researchers can accelerate the translation of scientific discoveries into tangible improvements in patient care, population health, and global health outcomes.

By the end of Chapter 11, readers will have a comprehensive understanding of the principles and processes involved in translational research. They will recognize the importance of bridging the gap between bench and bedside, and the potential impact of translational research in improving patient care and outcomes. This knowledge will empower researchers to engage in effective translational research practices, collaborate with stakeholders across the research continuum, and contribute to the development of innovative and impactful interventions.

12

FUTURE DIRECTIONS IN CLINICAL RESEARCH

CHAPTER XII

Chapter 12 explores the future directions in clinical research, highlighting the transformative potential of emerging trends and technologies. This chapter delves into advancements in technology and data science, precision medicine and genomics, patient-centered research, global health initiatives, and innovative research methodologies. By examining these future directions, the chapter provides insights into how clinical research can evolve to address the challenges and opportunities of the modern healthcare landscape. Readers will gain an understanding of the implications of these advancements for improving healthcare outcomes, fostering innovation, and ensuring ethical and socially responsible research practices. This knowledge will equip researchers to anticipate and navigate the future of clinical research, contributing to the ongoing progress and impact of their work.

SECTION 1: INTRODUCTION TO FUTURE DIRECTIONS

Introduction to Future Directions

In the first section of this chapter, we will explore the exciting possibilities and emerging trends in clinical research that are shaping the future of healthcare. Rapid advancements in technology, the focus on personalized medicine, and the growing importance of patient-centered research are driving transformative changes in the field. Let's discuss some of these future directions.

Advancements in Technology

Technology is revolutionizing clinical research, enabling new approaches to data collection, analysis, and healthcare delivery. For example, wearable devices, mobile health applications, and remote monitoring tools provide real-time data on patients' health status, allowing for continuous and personalized monitoring. Artificial intelligence (AI) and machine learning algorithms can analyze vast amounts of data to identify patterns, predict outcomes, and support clinical decision-making. Telemedicine and telehealth platforms facilitate remote consultations and improve access to care. These technological advancements have the potential to enhance research efficiency, patient engagement, and healthcare outcomes.

Personalized Medicine

Personalized medicine, also known as precision medicine, is an emerging approach that takes into account individual variability in genes, environment, and lifestyle to tailor healthcare interventions. Advances in genomics, molecular profiling, and biomarker identification are enabling targeted therapies and interventions based on an individual's unique characteristics. Genetic testing and genomic sequencing can guide treatment decisions, predict disease risk, and identify optimal drug therapies. Precision medicine holds promise for improving treatment outcomes, reducing adverse effects, and optimizing healthcare resource allocation.

Patient-Centered Research

Patient-centered research places patients at the center of the research process, actively involving them in study design, decision-making, and outcomes assessment. It recognizes the importance of patient preferences, values, and experiences in shaping healthcare interventions and policies. Patient-reported outcomes and patient engagement in research contribute to more relevant and meaningful research findings. Participatory research approaches, such as co-design and community-based research, empower patients and communities to actively participate in research activities. By incorporating patient perspectives, patient-centered research aims to improve treatment outcomes, patient satisfaction, and healthcare delivery.

These future directions in clinical research have significant implications for improving healthcare outcomes and transforming the way research is conducted. By harnessing technological advancements, embracing personalized medicine, and adopting patient-centered research approaches, researchers can enhance the precision, efficiency, and relevance of their studies.

SECTION 2: ADVANCEMENTS IN TECHNOLOGY AND DATA SCIENCE

Advancements in Technology and Data Science

In this section, we will delve into the impact of advancements in technology and data science on clinical research. The integration of artificial intelligence (AI), machine learning, and big data analytics is transforming the research landscape, enabling researchers to harness vast amounts of data and extract valuable insights. Let's explore the key areas where technology and data science are shaping clinical research.

Artificial Intelligence and Machine Learning

AI and machine learning algorithms have the ability to analyze complex datasets and identify patterns, relationships, and predictive models. In clinical research, AI can assist in tasks such as image analysis, diagnosis, risk prediction, and treatment recommendation. For example, deep learning algorithms have shown promise in interpreting medical images, such as radiographs and histopathology slides, with accuracy comparable to human experts. Machine learning models can also be trained to predict disease outcomes, treatment responses, and adverse events based on patient characteristics and clinical data. By leveraging AI and machine learning, researchers can gain new insights, improve decision-making, and enhance the efficiency of data analysis in clinical research.

Big Data Analytics

The proliferation of electronic health records (EHRs), genomic databases, and other large-scale data sources has led to the emergence of big data analytics in healthcare research. Big data analytics involves the processing and analysis of large and diverse datasets to uncover hidden patterns, correlations, and trends. Researchers can utilize big data to study population health, identify risk factors, and develop predictive models for disease prevention and management. Additionally, real-world data from sources like wearables, social media, and mobile health apps can provide valuable insights into patients' behaviors, preferences, and treatment outcomes. By leveraging big data analytics, clinical researchers can gain a deeper understanding of diseases, treatment effectiveness, and population health trends.

Technology-enabled Data Collection and Remote Monitoring

The widespread adoption of technology has facilitated data collection in real-world settings and enabled remote monitoring of patients. Electronic health records streamline data collection and provide researchers with access to comprehensive patient information. Wearable devices, such as fitness trackers and smartwatches, collect physiological data continuously, allowing for remote monitoring of vital signs, physical activity, and sleep patterns. Telemedicine platforms enable remote consultations, reducing geographical barriers and improving access to healthcare. These advancements in technology-driven data collection and remote monitoring provide researchers with rich and real-time data, enabling more accurate and comprehensive analysis.

While advancements in technology and data science offer tremendous opportunities in clinical research, they also come with challenges. Ensuring data privacy and security, addressing data quality issues, and promoting interoperability among different systems are important considerations. Additionally, researchers need to be mindful of potential biases and limitations associated with the use of AI and big data analytics. Collaborations between researchers, data scientists, and technology experts are essential to effectively integrate technology and data science into clinical research and overcome these challenges.

In summary, advancements in technology and data science are revolutionizing clinical research by enabling sophisticated data analysis, real-time data collection, and remote monitoring. AI, machine learning, and big data analytics have the potential to enhance research efficiency, improve diagnostic accuracy, and personalize healthcare interventions. As researchers embrace these technological advancements, they must also address the associated challenges and ensure ethical and responsible use of data.

SECTION 3: PRECISION MEDICINE AND GENOMICS

Precision Medicine and Genomics

In this section, we will explore the field of precision medicine and genomics and its implications for clinical research. Precision medicine aims to tailor medical treatments and interventions to individual patients based on their unique genetic makeup, environment, and lifestyle factors. Genomics, which refers to the study of an individual's entire DNA sequence and its variations, plays a crucial role in precision medicine. Let's delve into the key aspects of precision medicine and genomics.

Genetic and Genomic Information

Genetic and genomic information provides valuable insights into an individual's susceptibility to diseases, their response to treatments, and their likelihood of developing adverse drug reactions. Researchers are identifying and studying genetic variants associated with various diseases, allowing for personalized risk assessment and targeted interventions. For example, in oncology, genomic profiling of tumors can guide treatment decisions by identifying specific genetic alterations that can be targeted with precision therapies. In rare genetic diseases, identifying the underlying genetic mutation enables tailored treatments and improves patient outcomes.

Biomarkers and Genetic Testing

Biomarkers are measurable indicators, such as genetic variants, proteins, or metabolites, that can help predict disease risk, diagnose conditions, and monitor treatment responses. Genetic testing plays a significant role in identifying biomarkers associated with specific diseases. For instance, in cardiovascular research, genetic testing can identify individuals with genetic predispositions to conditions like familial hypercholesterolemia, enabling early intervention and personalized management strategies. Biomarkers and genetic testing facilitate the identification of patient subgroups that are more likely to benefit from certain treatments, leading to improved clinical outcomes.

Targeted Therapies

Precision medicine has led to the development of targeted therapies designed to act on specific molecular targets in diseases. By identifying genetic alterations or biomarkers that drive disease progression, researchers can develop therapies that directly address the underlying causes. Examples include targeted therapies for cancer, such as tyrosine kinase inhibitors that specifically inhibit abnormal signaling pathways in tumor cells. Targeted therapies often result in more effective treatment, reduced side effects, and improved patient outcomes.

While precision medicine and genomics hold great promise, several ethical considerations and challenges must be addressed. These include ensuring privacy and confidentiality of genomic data, addressing disparities in access to genetic testing and treatments, and managing the interpretation and communication of complex genetic information to patients and healthcare providers. There is also a need for robust evidence to support the clinical utility and cost-effectiveness of genomic testing and targeted therapies.

The future of precision medicine lies in integrating genomics with other 'omics' fields, such as proteomics and metabolomics, to gain a more comprehensive understanding of diseases and treatment responses. Advances in technologies, such as next-generation sequencing and high-throughput screening, are driving the rapid expansion of genomic data, enabling researchers to identify new therapeutic targets and develop novel interventions.

In conclusion, precision medicine and genomics have revolutionized clinical research by providing insights into personalized risk assessment, targeted therapies, and improved treatment outcomes. By harnessing genetic and genomic information, researchers can identify biomarkers, develop targeted therapies, and optimize patient care. However, ethical considerations, challenges in data interpretation, and access to genomic testing and treatments remain important areas for ongoing research and development.

SECTION 4: PATIENT-CENTERED RESEARCH AND PARTICIPATORY APPROACHES

Patient-Centered Research and Participatory Approaches

In this section, we will explore the shift towards patient-centered research and participatory approaches in clinical research. Traditional research has often been conducted with minimal input from patients and the public, leading to a disconnect between research findings and the needs and preferences of those directly affected by the research outcomes. Patient-centered research aims to address this gap by actively involving patients and the public in various stages of the research process. Let's continue the discussion into the key aspects of patient-centered research and participatory approaches.

Study Design and Prioritization

Patient-centered research starts by involving patients and the public in the identification and prioritization of research topics and questions. Engaging patients helps ensure that research addresses their real-life concerns, experiences, and outcomes that matter most to them. Patient advisory groups, community consultations, and patient-led organizations can provide valuable insights into the research agenda and help shape the study design.

Inclusion and Engagement

Patient-centered research emphasizes the inclusion and engagement of patients as active partners rather than passive subjects. Patients and the public can contribute their unique perspectives, knowledge, and lived experiences to enhance the research process. Researchers can involve patients in study design, data collection, analysis, and interpretation, promoting shared decision-making and co-learning. Involving patients as research team members, co-investigators, or research advisors can lead to more relevant and impactful research outcomes.

Patient-Reported Outcomes

Patient-centered research places a strong emphasis on incorporating patient-reported outcomes (PROs) into study measures. PROs capture the patient's perspective on their health status, symptoms, quality of life, and treatment preferences. By integrating PROs, researchers gain a more comprehensive understanding of the impact of interventions on patients' lives and can make more patient-centered treatment decisions.

Dissemination and Knowledge Translation

Patient-centered research aims to ensure that research findings are effectively communicated and disseminated to patients, healthcare providers, policymakers, and the broader public. Clear and accessible dissemination strategies, including plain language summaries and patient-friendly materials, enable patients to understand and apply research findings in their own healthcare decisions. Patient engagement in the dissemination process helps bridge the gap between research and practice and promotes the uptake of evidence-based interventions.

The shift towards patient-centered research and participatory approaches has several benefits. It enhances the relevance and applicability of research findings to patients' real-world experiences, leading to more patient-centered healthcare delivery. Involving patients in the research process empowers them as partners in their own care and promotes shared decision-making. Patient-centered research also strengthens trust between researchers, patients, and the public, fostering a collaborative approach to healthcare improvement.

However, there are challenges to consider when implementing patient-centered research. These include ensuring diverse and representative patient engagement, addressing power imbalances between researchers and patients, and providing adequate support and resources for meaningful patient involvement. Researchers must also navigate ethical considerations, such as informed consent and privacy protection, when involving patients in research activities.

In conclusion, patient-centered research and participatory approaches are transforming the landscape of clinical research by involving patients and the public as active partners in the research process. By incorporating patient perspectives, preferences, and values, research becomes more relevant, impactful, and patient centered. The integration of patient-reported outcomes and the meaningful dissemination of research findings further strengthen the connection between research and practice, ultimately improving healthcare outcomes and empowering patients in their healthcare journey.

SECTION 5: GLOBAL HEALTH AND CROSS-CULTURAL RESEARCH

Global Health and Cross-Cultural Research

In this section, we will delve into the field of global health and cross-cultural research, focusing on the importance of addressing health disparities, understanding cultural nuances, and conducting research in diverse populations. Global health research aims to improve health outcomes and reduce health disparities across different regions and populations worldwide. Let's explore the key aspects of global health research.

Health Disparities and Inequities

Global health research recognizes the existence of health disparities and inequities that exist both within and between countries. It aims to understand the social, economic, and environmental determinants of health and develop interventions that address these disparities. Researchers in this field work towards reducing disparities in access to healthcare, improving health outcomes, and promoting health equity globally.

Cultural Nuances and Contextual Factors

Conducting research in diverse populations requires an understanding of cultural nuances and contextual factors that influence health behaviors, beliefs, and outcomes. Cultural competence is crucial to ensure that research methods, interventions, and communication strategies are sensitive and appropriate for the cultural context. Researchers need to engage with local communities, respect cultural practices and beliefs, and adapt research approaches to fit the cultural context.

Ethical Considerations

Global health research raises unique ethical considerations, including informed consent, privacy protection, and power dynamics between researchers and communities. Researchers must navigate these ethical challenges by working closely with local ethics committees, respecting community values and norms, and ensuring that the research benefits outweigh any potential harms. Collaborations with local researchers and community stakeholders foster ethical research practices and community ownership.

Collaborations and Partnerships

Successful global health research relies on collaborations and partnerships between researchers, institutions, and communities across different countries and regions. Collaborations enhance the cultural sensitivity and relevance of research, facilitate knowledge exchange, and build local research capacity. Engaging local researchers and stakeholders in the research process fosters mutual learning, promotes sustainability, and facilitates the implementation of research findings.

Future Directions and Impact

The future of global health research holds promising opportunities for improving health outcomes worldwide. Advancements in technology, such as mobile health applications and telemedicine, enable the delivery of healthcare interventions in remote and underserved areas. Implementation research and health systems strengthening play a crucial role in translating research findings into sustainable healthcare practices. Additionally, addressing emerging global health challenges, such as infectious diseases, non-communicable diseases, and climate change, requires multidisciplinary collaborations and innovative approaches.

Examples of global health research include studies on infectious disease control and prevention, maternal and child health interventions, implementation of vaccination programs, and the impact of sociocultural factors on health behaviors. For instance, research conducted in low-income countries to improve access to clean water and sanitation can have a significant impact on reducing waterborne diseases and improving overall health outcomes. Similarly, research on cultural beliefs and practices related to healthcare-seeking behavior can inform the development of culturally appropriate interventions to address health disparities.

In conclusion, global health and cross-cultural research play a vital role in understanding and addressing health disparities and improving health outcomes globally. By recognizing and respecting cultural nuances, engaging with local communities, and promoting ethical research practices, researchers can contribute to reducing health inequalities and promoting health equity. Collaborations and partnerships between researchers, institutions, and communities are essential for the success of global health research. The future of this field holds immense potential for leveraging technology, strengthening health systems, and addressing emerging global health challenges to improve the well-being of populations worldwide.

SECTION 6: EMERGING RESEARCH METHODOLOGIES AND DESIGNS

Emerging Research Methodologies and Designs

In this section, we will explore emerging research methodologies and designs that are revolutionizing clinical research. These innovative approaches offer new ways to generate evidence, accelerate the translation of research findings into practice, and address the complex challenges in healthcare. Let's dive into some of these methodologies.

Adaptive Clinical Trial Designs

Adaptive clinical trials are flexible trial designs that allow modifications to key aspects of the study as data accumulates. These designs enable researchers to make interim decisions based on emerging data, such as sample size adjustments, treatment arm modifications, or population subgroups analyses. Adaptive designs increase trial efficiency, optimize resource allocation, and enhance the likelihood of identifying effective interventions. For example, the use of adaptive designs in oncology research has allowed for more efficient evaluation of novel therapies and identification of biomarkers associated with treatment response.

Real-World Evidence Studies

Real-world evidence (RWE) studies leverage data from routine clinical practice, electronic health records, claims databases, and other real-world sources to generate evidence on treatment effectiveness, safety, and comparative outcomes. RWE studies provide insights into how interventions perform in real-world settings, beyond the controlled environment of clinical trials. These studies can include large and diverse patient populations, making the findings more generalizable. RWE studies can complement traditional clinical trials by providing long-term safety and effectiveness data. For example, RWE studies have been instrumental in assessing the real-world effectiveness of new medications or comparing different treatment strategies in routine clinical practice.

Implementation Research

Implementation research focuses on understanding how to effectively implement evidence-based interventions into routine clinical practice. It investigates the barriers and facilitators to the adoption of new practices, identifies implementation strategies, and evaluates their impact on healthcare outcomes. Implementation research often utilizes mixed methods, combining qualitative and quantitative approaches, to provide a comprehensive understanding of the complex factors influencing implementation success. This type of research is critical for translating research findings into practice and ensuring their sustainable integration into healthcare systems. For example, implementation research has been applied to improve the adoption of evidence-based guidelines, enhance medication adherence, and implement quality improvement initiatives.

Advantages and Limitations

These emerging methodologies offer several advantages. Adaptive clinical trial designs provide flexibility, allowing researchers to refine study parameters in response to accumulating data, potentially reducing time and resource requirements. Real-world evidence studies provide insights into how interventions perform in real-world settings, enhancing generalizability and complementing controlled clinical trial data. Implementation research focuses on the practical application of research findings, ensuring effective translation into routine clinical practice. However, these methodologies also present challenges. Adaptive designs require careful planning, statistical expertise, and regulatory considerations to ensure scientific rigor and preserve patient safety. Real-world evidence studies face challenges related to data quality, data integration, and potential biases inherent in observational data. Implementation research requires collaboration with healthcare providers, policy makers, and other stakeholders to effectively integrate evidence-based practices into complex healthcare systems.

Regulatory Considerations

The adoption of these emerging methodologies necessitates regulatory considerations. Regulatory agencies, such as the FDA and EMA, are actively exploring ways to accommodate adaptive designs and real-world evidence in regulatory decision-making. Guidelines and frameworks are being developed to ensure the scientific validity, transparency, and reliability of data generated through these methodologies. Additionally, collaborations between regulatory agencies, researchers, and industry stakeholders facilitate the adoption of these methodologies while ensuring patient safety and maintaining high research standards.

Examples of these methodologies in action include adaptive clinical trials in precision medicine, where treatment arms are modified based on biomarker data, real-world evidence studies assessing the long-term effectiveness of vaccines in diverse populations, and implementation research evaluating the implementation strategies for evidence-based guidelines in primary care settings.

In conclusion, emerging research methodologies and designs, such as adaptive clinical trial designs, real-world evidence studies, and implementation research, offer innovative approaches to generate evidence, accelerate the translation of research findings into practice, and address the complexities of healthcare. These methodologies have the potential to enhance the efficiency and effectiveness of clinical research and improve patient outcomes. However, careful consideration of their advantages, limitations, regulatory requirements, and collaboration between stakeholders is crucial for their successful adoption and implementation.

SECTION 7: ADDRESSING ETHICAL AND SOCIAL IMPLICATIONS

Addressing Ethical and Social Implications

In this final section, we will explore the ethical and social implications of future directions in clinical research. As clinical research advances and embraces technologies like big data analytics and precision medicine, it is essential to address the ethical considerations and societal implications that arise. Let's delve into some of the key areas.

Ethical Frameworks and Privacy Protection

With the increasing utilization of big data and the integration of diverse data sources, protecting patient privacy and maintaining data confidentiality become paramount. Ethical frameworks, such as informed consent, data anonymization, and privacy regulations, are crucial for safeguarding individuals' rights and ensuring their trust in research. Researchers must adhere to strict ethical guidelines to ensure data privacy, security, and responsible use. For example, in genomic research, protecting the privacy of genetic information and providing clear consent processes for data sharing are critical considerations.

Equity, Access, and Social Determinants of Health

Future directions in clinical research should address issues of equity, access, and social determinants of health to ensure that advancements benefit all individuals, regardless of their socioeconomic status, race, ethnicity, or geographical location. Researchers must actively consider how new interventions, technologies, and research approaches can reduce health disparities and promote health equity. For instance, ensuring access to precision medicine interventions for underserved populations and considering social determinants of health in the design and implementation of research studies can lead to more equitable healthcare outcomes.

Data Sharing and Collaboration

The era of big data and precision medicine brings opportunities for enhanced collaboration and data sharing among researchers, institutions, and stakeholders. Open data sharing can accelerate research progress and enable the validation of findings. However, ethical considerations must be addressed, including ensuring data security, protecting patient privacy, and establishing guidelines for responsible data sharing. Encouraging collaborations and establishing data-sharing agreements that promote transparency, while respecting individual rights, can lead to significant advancements in research.

Role of Policymakers and Stakeholders

Policymakers play a crucial role in shaping the ethical and social landscape of future clinical research. They need to develop and enforce regulations and guidelines that promote ethical conduct, privacy protection, and equitable access to research outcomes. Stakeholder engagement, including patients, healthcare professionals, advocacy groups, and community representatives, is essential to ensure that research aligns with societal values and addresses the needs of diverse populations.

Examples of addressing ethical and social implications include the development of data governance frameworks that balance data sharing for research purposes while safeguarding privacy, initiatives to increase diversity in research participation, and policy interventions aimed at reducing health disparities through targeted research and interventions.

In conclusion, future directions in clinical research bring forth ethical and social considerations that must be carefully addressed. By establishing ethical frameworks, ensuring privacy protection, addressing equity and social determinants of health, promoting responsible data sharing, and engaging policymakers and stakeholders, we can navigate these challenges and advance research that is not only scientifically robust but also ethically sound and socially impactful. It is essential for researchers, policymakers, and stakeholders to collaborate and proactively address these ethical and social implications to ensure that future clinical research benefits individuals and communities alike.

Conclusion

By the end of Chapter 12, readers will have a comprehensive understanding of the exciting future directions in clinical research. They will recognize the transformative potential of advancements in technology, personalized medicine, patient-centered research, global health, emerging methodologies, and ethical considerations. This knowledge will empower researchers to embrace innovation, anticipate future trends, and contribute to the advancement of clinical research for the benefit of patient care and public health.

Made in the USA
Coppell, TX
05 November 2026

11307509R00292